W9-AHP-837

DISCARD

BETHANY
COLLEGE
LIBRARY

THE SEARCH
FOR
UNDERSTANDING

THE SEARCH FOR UNDERSTANDING

*Selected Writings of Scientists of the
Carnegie Institution, Published on the Sixty-Fifth
Anniversary of the Institution's Founding*

Edited by CARYL P. HASKINS

1967

CARNEGIE INSTITUTION
1530 P STREET, N.W., WASHINGTON, D. C.

All Rights Reserved
Including the Right of Reproduction
In Whole or in Part in Any Form
Except with Permission

Copyright 1967 by Carnegie Institution
of Washington, 1530 P Street, N.W.,
Washington, D. C.

Day unto day uttereth speech

and night unto night showeth knowledge . . .

There is no speech nor language

where their voice is not heard . . .

The heavens declare

the glory of God and the firmament

showeth his handiwork

— Psalm 19, 1–3, as quoted
in the frieze of the Elihu
Root Auditorium, Carnegie
Institution, Washington, D.C.

508.1
142735

/00 122

TABLE OF CONTENTS

viii

Remarks by
ANDREW CARNEGIE
On Presenting His Trust Deed

Mr. Chairman and Members of the Board of Trustees:

I beg first to thank you for so promptly and so cordially coming forward to aid me in this work by the acceptance of trusteeship. The President of the United States writes me in a note of congratulation, "I congratulate you especially upon the character, the extraordinarily high character, of the trustees." Those are his words. I believe that that estimate has been generally approved throughout the wide boundaries of our country.

May I say to you that my first idea while I dwelt upon the subject during the summer in Scotland was that it might be reserved for me to fulfil one of Washington's dearest wishes—to establish a university in Washington. I gave it careful study when I returned and was forced to the conclusion that if he were with us here today his finely balanced judgment would decide that such, under present conditions, would not be the best use of wealth. It was a tempting point suggested to me by the president of the women's George Washington Memorial Association, that the George Washington Memorial University, founded by Andrew Carnegie, would link my name with Washington. Well, perhaps that might justify such association with Washington, and perhaps it is reserved for some other man in the future to win that unique place; because if we continue to increase in population as we have done it is not an improbability that it may become a wise step to fulfil Washington's wish. But while that may justify the association of any other name with his, which is a matter of doubt, still I am very certain nothing else would. A suggestion that this gift of mine, which has its own field, which has nothing to do with the University, except as an aid to one if it is established, which has a field of its own, . . . is entitled to the great name of Washington, is one which I never for a moment could consider. If the coming university under the control of the Nation—as Washington suggested a national institution—is to be established, as it may be in the future, I think the name of Washington should be reserved for that and for that alone. Be it our opportunity in our day and generation to do what we can to extend the boundaries of human knowledge by utilizing existing institutions.

This is intended to cooperate with all existing institutions because one of the objections—the most serious one, which I could not overcome when I was desirous to establish a university here to carry out Washington's idea—was this: That it might tend to weaken existing institutions, while my desire was to cooperate with all kindred institutions, and to establish what would be a source of strength to all of them and not of weakness, and therefore I abandoned the idea of a Washington University or anything of a memorial character.

Gentlemen, a university worthy of Washington, or a memorial worthy of Washington, is not one costing a million dollars, or ten million dollars, or twenty million dollars, but of more. When I contemplated a university in Washington in fulfillment of Washington's great wish I set a larger amount than the largest of these. I take it for granted that no one or no association would think of using the revered name of Washington except for a university of first class rank, something greater and better, if I may be allowed to say so, than we have in our land today—and you all know the sums which are now used for our universities.

Gentlemen, your work now begins; your aims are high, you seek to expand known forces, to discover and utilize unknown forces for the benefit of man. Than this there can scarcely be a greater work. I wish you abundant success, and I venture to prophesy that through your efforts, in cooperation with kindred organizations, our country's contributions through research and the higher science in the domain of which we are now so woefully deficient, will compare in the near future not unfavorably with those of any other land.

Again, gentlemen, from my heart, I thank you, and I will now, with your permission read the deed of trust which has been prepared. I may say that the intended officers of this Institution have a letter from my cashier, stating that the notice of the transfer of the bonds will be sent you early in February. They cannot be transferred until the first of the month. They begin to bear interest on the first day of February. Here is the deed of trust.

There is nothing so important, I think, as the last clause. This clause follows the deed given to the Scotch universities, in the main. When I proposed it to the committee the chairman said he did not know about assuming so much responsibility as a trustee, and several gentlemen also suggested that it was too liberal, and threw too much responsibility upon them. Mr. Arthur Balfour was one of these. I replied to him that my experience was that it is not without the greatest difficulty we find men who can legislate for their own generation, and sometimes we are not quite successful even in that; "But," I asked, "Have you ever seen, or heard of a body of men wise enough to legislate for the next generation?" He answered, "No, I never have"; and "You are quite right; that is the wisest provision I have ever heard of in a trust deed."

I have nothing more to say to you, gentlemen, having already expressed my thanks; but, as I began with doing this, I feel that I should also like to end doing so, and therefore, I thank you again.

INTRODUCTION

This year marks the sixty-fifth anniversary of the Carnegie Institution of Washington. Sixty-five years ago Andrew Carnegie transmitted to a newly elected Board of Trustees a deed conveying the sum of ten million dollars "to found, in the city of Washington, an Institution which with the cooperation of institutions now or hereafter established, there or elsewhere, shall in the broadest and most liberal manner encourage investigation, research, and discovery. . . ." At the end of January of 1902, the Trustees elected Daniel Coit Gilman, fresh from his notable career as the president of the Johns Hopkins University, as the first president of the Carnegie Institution, and resolved "to promote original research by systematically sustaining projects of broad scope that may lead to the discovery and utilization of new forces for the benefit of man . . . projects of minor scope that may fill in gaps of knowledge of particular things or restricted fields of research . . . administration of a definite or stated research under a single direction by competent individuals."

It was not the first of Andrew Carnegie's great philanthropic gifts. Far from it indeed. In the last decade of the closing century he had established in Pittsburgh the Carnegie Institute with its natural history museum, its music hall, and its department of fine arts, and had made possible the Carnegie Institute of Technology, grown now to high rank among the scientific and technical universities of the nation. In the opening years of the new century he had established the Carnegie Trust for the Universities of Scotland, and the Carnegie Dunfermline Trust for the benefit of his native town. Nor was the Carnegie Institution, by many removes, to be the last of his gifts. It was followed by the Carnegie Foundation for the Advancement of Teaching, the Carnegie Endowment for International Peace, Carnegie Hero Funds in no less than eleven countries, and in culmination, the Carnegie Corporation of New York.

But the establishment of the Carnegie Institution of Washington marked a wholly new direction in Mr. Carnegie's benefaction. More than that, it marked the inception of a wholly new kind of institution in American life—the first to be devoted wholly and completely, in intent and in philosophy, to the ideal of research scholarship over wide fronts of science in its broadest, most unfettered, most completely uncommitted aspects. This was a novel concept indeed and some of the records of the time leave no doubt that, like most great and novel ideas and ventures, it was not everywhere acceptable— nor indeed always comprehensible to a young nation with a strongly established tradition of pragmatism.

Four years after its establishment, the Institution had been granted a new Charter by special Act of Congress, and had been organized into no less than fourteen departments, representing as many stated areas of study. Over the next five years, several definitive judgments were made as to how and where the Institution could work most effectively. One judgment taken during these years of experiment was to shape its future history decisively. This was the decision to concentrate the resources of the Institution primarily in the research of its various departments: to make of it, in essence, an operating rather than a granting scientific organization. By 1911, its endowment more than doubled by subsequent grants from Mr. Carnegie, its departments firmly established but now reduced to ten, the Institution was molded to the purpose, and had taken on essentially the form of organization, that characterize it today. Through the following years new departments have arisen, departments have been consolidated, and some departments have been closed, as the needs and the research frontiers of each changing decade have dictated. Whole fields that were represented in the Institution in 1911, such as economics and sociology, historical research, meridian astrometry, nutrition in the purely medical sense, no longer are included in its program, as the resources of the nation in those areas have strengthened and enlarged. Other fields not even existent then but now on the very frontiers of modern research, like modern embryology, molecular and cellular biology, the study of the mechanisms of photosynthesis, have been embraced within its program in more recent years.

Today there are five, instead of ten, departments in the Institution. Most originated in planning going back for a considerable time, though the work they conduct today has expanded far beyond the boundaries originally conceived. The Department of Terrestrial Magnetism was founded in 1904, the Geophysical Laboratory in 1906. Both carry on their research today in Washington at the very frontiers of their respective fields. A Desert Laboratory, later to become the Division and then the Department of Plant Biology, appeared in 1903. The Department of Plant Biology, today one of the leading laboratories in the nation in research into the mechanisms of photosynthesis in green plants at cellular and molecular levels, and in certain aspects of the plant evolution, continues its work on the campus of Stanford University in Palo Alto. A Solar Observatory on Mount Wilson was planned as early as 1902. Studies of the sun remain at the pioneering fringes of investigation in that part of the Institution to this day, and currently they are being greatly extended. But now the original Solar Observatories have metamorphosed to the complex of giant telescopes—including the largest in the world—of the Mount Wilson and Palomar Observatories, operated by the Institution jointly with the California Institute of Technology. To the intensive program of solar investigations of which George Ellery Hale dreamed, and which he initiated with his striking discoveries of magnetic fields in the sun, have been added a goodly share of the world's most important findings about the farthest reaches of the celestial universe, including, in the last years, the momentous discovery, and the subsequent investigation, of the quasi-stellar energy sources of the sky.

As early as 1904 a Station for Experimental Evolution had been established by the Institution at Cold Spring Harbor on Long Island, and here it was that George Harrison Shull, as he described in his paper, "A Pure-Line Method in Corn Breeding," presented in this volume, conceived the general principles underlying the development of hybrid maize, providing the scientific basis for an innovation in plant agronomy that by 1952 was estimated to have brought an economic gain for the United States of almost forty billion dollars. Through all the years since Shull's work the Station for Experimental Evolution, which was to become the Department of Genetics and now is the Genetics Research Unit of the Carnegie Institution, has continued to pioneer over a wide range at the very frontiers of genetics research. Here was first demonstrated the astonishing fact that some genes function primarily as regulators of others in the system of inheritance—a discovery which of recent years has had the most important implications for embryology and studies of human and animal development, as well as for the entire range of heredity. Here the whole field of viral genetics was pioneered, and important steps taken, which are presently being extended, in the understanding of the details of the marvelously precise functioning of DNA in inheritance and development.

At Baltimore, in 1914, a Department of Embryology was established. First for long years as a guest in the Johns Hopkins Medical School, and most recently in its own laboratory on the Homewood Campus of Johns Hopkins in Baltimore, the Department has consistently maintained its position as one of the greatest research organizations in the world in the field of embryonic development taken in its widest sense. In its early years particularly, many of its concerns were quite directly related to clinical fields. This is still true to some extent. But with the years the program of the Department has become more and more fundamental, more and more deeply concerned with the determining events of embryogenesis at cellular and subcellular levels.

The philosophy that lay behind the gift of Mr. Carnegie, and that shaped the course of the Institution, was not only notable and original, but most enduring, as six following decades amply demonstrated. Through all the years, the major philosophies of the Institution and one major feature of its organizational pattern have stood constant, tested and re-tested in situation after situation and proved repeatedly as fresh and relevant today as when they were conceived. The decision made at the outset that flexibility and effectiveness in the kind of research to which the Institution is dedicated can best be achieved through a group of rather small unit laboratories, each mobile and relatively independent, each able to seize the initiative in new fields as they appear, yet all sufficiently related so that they may be of mutual assistance when opportunity dictates, was a remarkably prescient one. Over the succeeding decades, as research has burgeoned in the nation and organizations devoted to research have multiplied in both the private and the public sectors, many other experiments in organizational form have been tried. But it is particularly noteworthy that some of the most modern thinking and experiment in the organization of research, both in this country and abroad, has returned to precisely this pattern as one of the most effective in exploring the dynamic leading edges of scientific knowledge.

Organization, however, is simply a framework, vital to be sure, but only supporting in its function. Most significant—

and most truly enduring—have been the elements of philosophy and purpose which inaugurated the Institution and which have remained unchanged through the years: the philosophy that all its resources, all its deepest purposes, are centered in the creative individual, whatever be his field, that in the truest sense he is the uncommitted investigator, suitably endowed and suitably protected, whose time, quite literally, is bought by the Institution and then returned as unconstrained endowment. And with this goes the philosophy, equally deep-seated and equally important, that this freedom from fixed commitment applies to fields of endeavor as well as to men: that high mobility within specific fields, that the unfettered crossing of fields, that the fashioning of unconventionally wide-ranging research programs, are subject only to the limitations imposed by Nature and by the judgments of gifted and discriminating investigators, and that making this mobility and flexibility possible is a principal objective of the Institution.

Over the years that philosophy, and the programs that have followed from it, have led to many pioneering practical discoveries within the Institution. In addition to the discovery of hybrid corn, the development of a strain of the mold *Penicillium,* producing three to five times as much penicillin as the highest yielding strains then known, was pioneered at the Department of Genetics during World War II when the antibiotic itself, still quite rare and immensely expensive, was vitally needed on the battlefields. In 1935, fully fifteen years before the extensive research on radar for combat in the Second World War, Breit and Tuve at the Department of Terrestrial Magnetism produced radio pulses and for the first time observed their echo from the ionosphere. At the Geophysical Laboratory, Day and Shepherd early undertook experiments in the field of low-expansion quartz glasses that proved basic to the evolution of Pyrex—a program that during the First World War supplied the United States with ninety-seven per cent of its requirements for optical glass. In the same laboratory, much later, studies by Morey on lanthanum and borate glasses of high refractive index resulted in the development of a whole new family of glasses of great importance in the manufacture of photographic lenses—a development having major implications for the United States in the Second World War. In the Geophysical Laboratory, again, Rankin and Wright as early as 1915 were able to solve the age-old riddle of cement, and their classic work has served ever since as a guide to the chemical aspects of the cement industry. From the same laboratory in later years have come new refractories for the steel industry, studies of natural geothermometers and geochronometers of fundamental concern to practical mining and oil prospecting as much as to fundamental geology.

Yet these practical contributions from the work of the Institution, useful and indeed vital as many of them have been over the years of its work, in one sense represent only by-products, mere projecting iceberg-tips, as it were, of the original and continuing vision, indicators only of the seven-eighths submerged. That seven-eighths, of course, lies in the kingdom of the mind. It lies in that devotion to deeper patterns, that determination to explore the symmetries, the lights and shades of Nature, wherever the search may lead, to which the Institution was originally dedicated and which, undeviatingly, it pursues today.

Striking innovations have occurred within that kingdom of the mind, and these, possibly in a truer sense than the sometimes more spectacular practical "firsts," stand as the proper signatures of the Institution. They range over many fields. While the thinking that underlay the famous Michelson-Morley experiment on "ether-drift" was still fresh, for example, Professor Michelson, holder of the first Nobel prize to be awarded in the United States in the natural sciences, repeated the experiment within the Institution with an accuracy hitherto unattained, giving strong support to the theory of relativity, itself still at a stage of question and doubt. Within the Institution too, Michelson repeated with great refinement the pioneering determination of the velocity of light for which he is perhaps best known. At the Mount Wilson and Palomar Observatories Hale's pioneering discovery that sunspots mark strong magnetic fields has been followed in more recent years by studies of solar magnetism of unprecedented refinement, and by the discovery, among the stars, of the most intense magnetic fields ever observed in any astronomical body. Hubble's studies of the phenomena of the redshift in stellar spectra led to the theory of the expanding universe, which has culminated in the last years in a drama of celestial discovery unrivaled since the days of Galileo—the discovery and the beginning of investigation of the nature of the quasi-stellar sources of the sky, apparently lying at distances in the universe undreamed of even a decade ago. The last years, too, have brought from the Observatories concepts of stellar ages, and of the courses of stellar evolution in the universe, that less than a quarter of a century ago would have been inconceivable.

At the Department of Terrestrial Magnetism, a series of conferences held shortly before the Second World War in cooperation with the George Washington University stimulated among other things the suggestion that the source of energy in the sun and stars is a nuclear reaction involving carbon—a notion which within a year led to the classical model of the hydrogen-helium transformation now familiar as one of the accepted sources of stellar energy. In the Geophysical Laboratory studies of the biochemistry of ancient sediments have given new dimensions to our concepts of the age of terrestrial life, which studies of the artificial synthesis of amino acids from inorganic chemical components under a variety of physical and chemical conditions, beside shedding light on the probable modes of the origin of life on earth and the nature of its environments, have also carried important theoretical implications for our notions about the existence of life on other planets. At the Department of Plant Biology, work on photosynthesis has produced suggestive insights about that critical step which, with all the research that has been brought to bear upon this outstandingly important subject over the last half century, still eludes our understanding—the initial process by which the energy of light is used in the fixation of carbon dioxide. In three Departments of the Institution: the Department of Terrestrial Magnetism, the Department of Embryology, and the Genetics Unit at Cold Spring Harbor, investigations of cellular metabolism and development, of cellular differentiation, and of the mechanisms of heredity have brought striking new knowledge of the detailed ways in which the materials of heredity and development interact at the level of the cell nucleus and its cytoplasm, at the level of the germinal cell and the body cell of the plant or animal, and at the level of differentiation and development of the individual organism.

These are but meager and scattered examples taken from the rich matrix of sixty-five years of Institution work. But they are a fair example of its most typical fruit—the truest product of the philosophy in which it was founded and through which it lives.

All scientific work of the kind carried forward in the Institution involves an important by-product—the vehicle through which it is first announced to the world, and which makes it permanent for later generations. If the investigation is of practical nature, that vehicle may be the patent, or the working model. If it is of more theoretical cast, the product is likely to be the book or the paper, specialized or generalized, narrow or broad, dry and formal or vibrant and classical in accord with the subject matter and the abilities and temperament and inclination of the writer. If the work is of philosophical cast, and if the writer, in addition to being a first-rate scientist, is also a first-rate man of letters, then there may be a rare by-product which can constitute one of the most enduring heritages of all for our culture, the brilliant scientific essay.

This particular by-product is indeed uncommon. It is restricted by the exigencies of time, and by limitations of practicability and interest and particular talent. But it has appeared within the Institution with unusual frequency. Perhaps there is something about the atmosphere of the Institution itself, and the collective temperament of those who work within it, that favors its creation.

It is the belief that this particular kind of contribution —the contribution to the fine literature of science—has a specially enduring value, that the present collection has been undertaken. It represents a first attempt to bring together a few of the more general accounts that have appeared from time to time over the years from the pens of members of the Institution—undertaken in the widest variety of contexts and for the widest variety of reasons.

The title of this book is drawn from an essay by Dr. Vannevar Bush published in 1952. Dr. Bush was President of the Carnegie Institution from 1939 to 1956. He was the architect of the fundamental ideas and designs upon which the whole development of modern computers was largely based. During the years of World War II, as Director of the Office of Scientific Research and Development, he was more completely responsible than any other individual in the nation for the mobilization and direction of the massive national scientific and technical effort that proved critical to victory— and indeed to survival—in that conflict, and which has since become nearly legendary. Later, in his classic study summarized in the book, *Science the Endless Frontier,* Dr. Bush laid the foundation from which, in later years, the modern National Science Foundation was to be developed. Indeed, the whole current philosophy of federal aid to nongovernmental, scientific research in the nation virtually took its origin in the debates and deliberations of those years, guided and usually dominated by his thinking.

But in addition to all this, Dr. Bush was, and is, an inspired essayist. To him in special measure this book is dedicated. It opens with another—and one of the earliest—of his essays. "The Builders" appeared in the *Technology Review* of the Massachusetts Institute of Technology in January of 1945. That essay is a classic now, often reprinted, and widely familiar. Yet we have been able to find nothing that so vividly sets out some of the deepest and most abiding truths of science: truths that we need particularly to bear in mind today. Nowhere is it more poignantly evident that the great messages, in science as elsewhere, deserve continual rereading.

A special tribute goes to the Directors and the Staff of the Carnegie Institution in the making of this book. The original collection of essays from which the choices were made—more than four hundred in all—was derived largely from their nominations. This initial group was slowly sifted and reduced to approximately fifty, and these, in turn, were further winnowed to the twenty-two included here.

The criteria of choice were necessarily specialized. Vividness and that elusive quality, hardly to be defined, that makes the fine essay, were of course particularly sought. Of equal weight—and indeed closely linked—was the requirement that the subject itself should be exciting, and that the work described should have been, in its time, sufficiently pioneering and critical to the advance of the branch of science that it depicts to fully merit the quality of its description. Thus each essay, it is hoped, represents at once an example of fine writing in the service of science, and an exposition of real importance in the history of science.

Other necessary criteria, however, were more specialized. Thus it seemed desirable that the general balance of subjects treated approximate the balance of scientific concerns within the Carnegie Institution. This meant, inevitably and often heartbreakingly, that some of the contributions of the highest order had to be excluded only for the reason that they lay in subject areas already well represented and so could have been included only at the price of an undue sacrifice of proportion.

It will be evident that the task of selection was not easy. And though always rewarding, it was not always pleasant. For many essays from members of the Institution that, finally, were not incorporated surely merited inclusion just as fully as those that are represented here on every score save those imposed by these arbitrary strictures of space and balance.

But it is our hope that this collection, as it stands, will bring to the reader some idea of the scope and shape of the scientific interest of the Carnegie Institution. More, we hope that it may stand as one example of that partnership between fine science and fine writing upon which the best in scientific communication is so often based: one example of that welding of scientific and of humanistic endeavor that can provide our firmest guarantees that over future years, whatever the forces that tend to part them—real forces that we constantly combat and of which we must be perpetually aware—the deepest and the finest concerns of science and of the humanities will not be sundered.

CARYL P. HASKINS

ACKNOWLEDGMENTS

Again I want to express particular gratitude to the Directors of the Departments and all the members of the Staff of the Carnegie Institution for their selection and sifting of essays —contributions that have been critical to the whole endeavor. Equal appreciation goes to the contributors who have made their own creations available. To the publishers and the publications that have permitted the use of material already presented in print very special thanks are due. They are individually acknowledged in connection with their specific contributions.

This work has been, in the truest sense, a joint effort of many people. The final decisions, however, have in every case been those of the editor, and it is with him, and with him alone, that the responsibility for shortcomings must rest.

C. P. H.

I
THE SEARCH

*L*ate *in 1944 Dr. Vannevar Bush, then President of Carnegie Institution, was invited to write a short article for* Technology Review, *a publication of the Massachusetts Institute of Technology. The result was "The Builders," which appeared in* Technology Review *in January 1945.*

A distinguished scientist, an educator, an inventor, an accomplished administrator, Dr. Bush has had a career of superlatives. Born in 1890, in Everett, Massachusetts, he did his undergraduate work at Tufts University. In 1916 he was awarded a doctorate in engineering from Harvard University and Massachusetts Institute of Technology simultaneously. He taught at M.I.T. for 19 years, first as associate professor and then as professor of electric power transmission. In 1932 he became Dean of the School of Engineering and Vice President of the Institute.

He became President of the Carnegie Institution in 1938, bringing to his new position the skill of a trained investigator, the forcefulness of a leader, and the experience of an executive. In 1940, in the midst of his activities at the Institution, he took on additional responsibilities in connection with the war effort. President Roosevelt in 1941 appointed him Director of the newly established Office of Scientific Research and Development, which he administered from his office at the Carnegie Institution, mobilizing the scientific effort of the country, initiating programs of research in cooperation with the Army and Navy, and advising the President as to the status of defense-oriented research and development. Dr. Bush was thus a central figure in such wartime developments as nuclear fission, and shared responsibility for the determination of scientific policy throughout the war.

He retired as President of the Carnegie Institution in 1955, and is presently a Trustee. He continues to serve as an advisor to many organizations, as well as to the Federal Government.

Vannevar Bush

THE BUILDERS*

From *Technology Review,* January 1945. By permission of the
Alumni Association of the Massachusetts Institute of Technology.

The process by which the boundaries of knowledge
are advanced, and the structure of organized science is built,
is a complex process indeed. It corresponds fairly well with the
exploitation of a difficult quarry for its building materials and
the fitting of these into an edifice; but there are very significant
differences. First, the material itself is exceedingly varied,
hidden and overlaid with relatively worthless rubble, and the
process of uncovering new facts and relationships has some of
the attributes of prospecting and exploration rather than of
mining or quarrying. Second, the whole effort is highly unorganized. There are no direct orders from architect or quarry-
master. Individuals and small bands proceed about their busi-
nesses unimpeded and uncontrolled, digging where they will,
working over their material, and tucking it into place in the
edifice.

Finally, the edifice itself has a remarkable property,
for its form is predestined by the laws of logic and the nature
of human reasoning. It is almost as though it had once existed,
and its building blocks had then been scattered, hidden, and
buried, each with its unique form retained so that it would fit
only in its own peculiar position, and with the concomitant
limitation that the blocks cannot be found or recognized until
the building of the structure has progressed to the point where
their position and form reveals itself to the discerning eye of
the talented worker in the quarry. Parts of the edifice are being

* "The Builders" will also be included in a volume soon to be published by William Morrow and Company.

used while construction proceeds, by reason of the applications of science, but other parts are merely admired for their beauty and symmetry, and their possible utility is not in question.

In these circumstances it is not at all strange that the workers sometimes proceed in erratic ways. There are those who are quite content, given a few tools, to dig away unearthing odd blocks, piling them up in the view of fellow workers, and apparently not caring whether they fit anywhere or not. Unfortunately there are also those who watch carefully until some industrious group digs out a particularly ornamental block, whereupon they fit it in place with much gusto and bow to the crowd. Some groups do not dig at all, but spend all their time arguing as to the exact arrangement of a cornice or an abutment. Some spend all their days trying to pull down a block or two that a rival has put in place. Some, indeed, neither dig nor argue, but go along with the crowd, scratch here and there, and enjoy the scenery. Some sit by and give advice, and some just sit.

On the other hand there are those men of rare vision, who can grasp well in advance just the block that is needed for rapid advance on a section of the edifice to be possible, who can tell by some subtle sense where it will be found, and who have an uncanny skill in cleaning away dross and bringing it surely into the light. These are the master workmen. For each of them there can well be many of lesser stature who chip and delve, industriously, but with little grasp of what it is all about, and who nevertheless make the great steps possible.

There are those who can give the structure meaning, who can trace its evolution from early times, and describe the glories that are to be, in ways that inspire those who work and those who enjoy. They bring the inspiration that all is not mere building of monotonous walls, and that there is architecture even though the architect is not seen to guide and order.

There are those who labor to make the utility of the structure real, to cause it to give shelter to the multitude, that

5   *The Builders*

they may be better protected, and that they may derive health and well-being because of its presence.

And the edifice is not built by the quarrymen and the masons alone. There are those who bring them food during their labors, and cooling drink when the days are warm, who sing to them and place flowers on the little walls that have grown with the years.

There are also the old men, whose days of vigorous building are done, whose eyes are too dim to see the details of the arch or the needed form of its keystone; but who have built a wall here and there, and lived long in the edifice, who have learned to love it and who have even grasped a suggestion of its ultimate meaning; and who sit in the shade and encourage the young men.

The passion of George Sarton for knowledge and understanding of the origins, nature, philosophy and growth of science led him, throughout a long and busy life, to dig out the records of great discoveries and great scientists of the past, and to write voluminously, entertainingly, and with style and enlightenment about the history of this subject.

He was born in Ghent, Belgium, in 1884, and attended the University of Ghent, where he received the B.Sc. degree in 1906 and the Sc.D. in 1911. He came to the United States in 1915, to lecture on the history of science at George Washington University, and later at Harvard and Radcliffe. He became a citizen of the United States in 1924. He was professor of the history of science at Harvard from 1940 until 1951, and he was an Associate of the Carnegie Institution for thirty years, from 1918 until his retirement in 1948. He died in 1956 at his home in Cambridge, Massachusetts.

A perceptive editor and prolific writer, he not only produced numerous books and articles on the history of science, but in 1912 he founded and for many years edited Isis, *an international review devoted to the history and philosophy of science. In 1936 he founded and edited* Osiris, *a publication devoted to studies on the history of science, learning and culture. Perhaps his most important published work is his exhaustive three-volume* Introduction to the History of Science, *which traced the development of what is now called "the scientific method" from the beginnings of technology through the fourteenth century. This monumental 2090-page work was supported by the Carnegie Institution. Dr. Sarton also completed for the Institution a study of Leonardo da Vinci considered as a scientist, which is reprinted in this book, from his* The Life of Science, *published in 1948.*

Throughout his career Dr. Sarton insisted that the history of science is not the sum of the histories of the separate sciences, but rather their integration; he frequently referred to the history of science as a new discipline. The seeds he thus planted took root: there are now chairs of the history of science, and courses in the subject, at many leading universities, and scholars the world over consult his publications.

6

George Sarton

LEONARDO AND THE BIRTH OF MODERN SCIENCE

From *The Life of Science,* published by Henry Schuman, 1948.

Leonardo da Vinci died in the little manor of Cloux, near Amboise, where he had been for the previous three years the honored guest of Francis I, on May 2, 1519. He was not only one of the greatest artists, but even more the greatest scientist and the greatest engineer of his day. Indeed, with the passing of time his unique personality looms larger and larger and bids fair to attain, as soon as it is completely known, gigantic proportions.

Leonardo the artist is so well known that I shall hardly speak of him, but it is worth while for the purpose that I have in mind to recall briefly the most important facts of his life.

He was born in Vinci, a village in the hills between Florence and Pisa, in 1452, an illegitimate child, his mother being a peasant woman, and his father Ser Piero, a notary, a man of substance. The latter's family can be traced back to 1339, through three other generations of notaries. Soon after Leonardo's birth, his father took him away from his mother, and both parents hastened to marry, each in his own set. Ser Piero must have been a man of tremendous vitality, mental and physical. He was one of the most successful notaries of the Signoria and of the great families of Florence, and his wealth increased apace. He married four times, the two first unions remaining childless. His first legitimate child was not born until

7

1476, when Leonardo was already twenty-four, but after that ten more children were born to him by his third and fourth wives, the last one in the very year of his death, which occurred in 1504, when he was seventy-seven.

Thus Leonardo had five mothers. The real one disappears soon after his birth; she bore him and her mission ended there as far as Leonardo was concerned. What the four others were to him, we do not know, for he does not speak of them. He had five mothers and he had none. He is a motherless child, also a brotherless one, because he does not seem to have had much to do with his eleven brothers and sisters—far younger than himself anyhow—except when, at their father's death, they all leagued themselves against him to deny him any part of the patrimony. A motherless, brotherless, lonely childhood; we cannot lay too much stress on this; it accounts for so much.

In or about 1470 Ser Piero placed his son, now a very handsome and precocious boy, in the studio of Andrea Verrocchio, who since Donatello's death was the greatest sculptor of Florence; also a painter, a goldsmith, a very versatile man, indeed. Within the next years Leonardo had the opportunity to show the stuff of which he was made, and by 1480 his genius had matured. He was considered by common consent a great painter, and, moreover, his mind was swarming with ideas, not simply artistic ideas, but also architectural and engineering plans.

Leonardo was born in the neighborhood of Florence and bred in the great city. It is well, even in so short a sketch, to say what this implies. The people of Tuscany are made up of an extraordinary mixture of Etruscan, Roman, and Teutonic blood. Their main city, Florence, had been for centuries a considerable emporium, but also a center of arts and of letters. Suffice it to remember that of all the Italian dialects it is the Tuscan, and more specifically its Florentine variety, which has become the national language. The prosperous city soon took a lively interest in art, but loved it in its own way. These imaginative but cool-headed merchants patronize goldsmiths, sculptors, draftsmen. They do not waste any sentimentality, neither are they very sensual: clear outlines appeal more to them than gorgeous colors. Except when they are temporarily maddened by personal jealousy or by a feud which spreads like oil, it would be difficult to find people more level-headed, and having on an average more common sense and a clearer will.

Leonardo was a Florentine to the backbone, and yet this environment was not congenial to him. He was distinctly superior to most of his fellow citizens as a craftsman, but he could not match the best of them in literary matters. The Medici had gathered around them a circle of men whose delight it was to discuss topics of Greek, Latin, and vernacular literature, and to debate, often in a very learned manner, the subject of Platonic philosophy. There is no gainsaying that these Neoplatonists were a brilliant set of men, but their interests were chiefly of the literary kind; they were men of letters and loved beautiful discourse for its own sake. On the contrary, young Leonardo, following an irresistible trend, was carrying on scientific and technical investigations of every sort. The engineer in him was slowly developing. Perhaps, he could not help considering these amateur philosophers as idle talkers; but it is just as likely that, being a motherless child, he was not endowed with sufficient urbanity to fare comfortably in this society of refined dilettanti. Nature more and more engrossed his attention, and he was far more deeply concerned in solving its innumerable problems than in trying to reconcile Platonism and Christianity. Neither could his brother artists satisfy his intellectual needs; they were talking shop and fretting all the time. A few had shown some interest in scientific matters, but on the whole their horizon was too narrow and their self-centeredness unbearable. Also, Florence was becoming a very old place, and an overgrowth of traditions and conventions gradually crowded out all initiative and real originality. So Leonardo left and went to Milan, to the court of Ludovico

Sforza, at that time one of the most splendid courts of Europe. Milan would certainly offer more opportunities to an enterprising and restless mind like his. The very desire of outdoing Florence was a tremendous impulse for Ludovico: he was anxious to make of his capital a new Athens, and of the nearby university town of Pavia a great cultural center. His happiest thought perhaps was to keep around him two men who were among the greatest of their day—Bramante and Leonardo. The liberal opportunities which were offered to these two giants are the supreme glory of the Sforzas and of Milan.

Leonardo was employed by the Duke as a civil and military engineer, as a pageant master, as a sculptor, as a painter, as an architect. How far he was understood by his patron it is difficult to say. But he seems to have thrived in this new atmosphere, and these Milanese years are among the most active and the most fertile of his life. He was now at the height of his power and full scope was given to his devouring activity. It is during this period, for instance, that he modelled his famous equestrian statue of Francesco Sforza, that he painted the "Virgin of the Rocks," and the "Last Supper," while he was also superintending important hydraulic works, and pursuing indefatigably his various scientific investigations. Yet even at this time of greatest activity and enthusiasm he must have been a lonesome man. This brilliant but very corrupt court was of course the rendezvous of hundreds of dilettanti, parasites, snobs—male and female—and what could Leonardo do to protect himself against them but be silent and withdraw into his own shell?

Milan justly shares with Florence the fame of having given Leonardo to the world; it was really his second birthplace. Unfortunately, before long, heavy clouds gathered over this joyous city, and by 1500 the show was over and Ludovico, made prisoner by the French, was to spend the last ten years of his life most miserably in the underground cell of a dungeon. From that time on, Leonardo's life became very unsettled. It is true, he spent many years in Florence, employed by the Signoria, painting "la Gioconda" and the "Battle of Anghiari"; then for some years he was back in Milan, but he is more and more restless and somehow the charm is broken. After the fall of the Sforzas, Isabella d'Este, Marchioness of Mantua—perhaps the most distinguished woman of the Renaissance—tried to attach Leonardo to her service, but he refused, and instead he chose, in 1502, to follow Cesare Borgia as his military engineer. One may wonder at this choice, yet it is easy enough to explain. At that time Leonardo was already far prouder of his achievements as a mechanic and an engineer than as a painter. It is likely that in the eyes of Isabella, however, he was simply an artist and he may have feared that this accomplished princess would give him but little scope for his engineering designs and his scientific research. On the other hand, Leonardo found himself less and less at home in Florence. The city had considerably changed in the last ten years. Savonarola had ruled it, and many of the artists had been deeply swayed by his passionate appeals, and even more by his death. For once, fair Florence had lost her head. And then also, young Michael Angelo had appeared, heroic but intolerant and immoderate: he and Leonardo were equally great but so different that they could not possibly get on together.

In 1513–1515 Leonardo went to the papal court, but there, for the first time in his life, the old man was snubbed. Having left Rome, his prospects were getting darker, when fortunately he met in Bologna the young King of France, Francis I, who persuaded him to accept his patronage. The King offered him a little castle in Touraine, with a princely income, and there Leonardo spent in comparative quietness, the last three years of his life. It must be said to the credit of Francis I that he seems to have understood his guest, or at least to have divined his sterling worth. France, however, did not appreciate Leonardo, and was not faithful to her trust. The cloister of Saint-Florentin at Amboise, where the great artist had been buried, was destroyed by a fire in 1808, and his very ashes are lost.

He was apparently an old man when he died, much older than his years, exhausted by his relentless mind and by the vicissitudes and the miseries of his strange career. Only those who have known suffering and anxiety can fully understand the drama and the beauty of his life.

Throughout his existence Leonardo had carried on simultaneously, and almost without a break, his work as an artist, as a scientist, as an engineer. Such a diversity of gifts was not as unusual in his day as it would be now. Paolo Uccello, Leo B. Alberti, Piero dei Franceschi, even Verrocchio himself, had shown more than a casual interest in scientific matters such as perspective and anatomy, but Leonardo towers far above them. The excellence of his endowment is far more amazing than its complexity. His curiosity was universal to such a degree that to write a complete study of his genius amounts to writing a real encyclopedia of fifteenth-century science and technology. From his earliest age he had given proofs of this insatiable thirst for knowledge. He could take nothing for granted. Everything that he saw, either in the fields or on the moving surface of a river, or in the sky, or in the bottega of his master, or in the workshops of Florence, raised a new problem in his mind. Most of the time neither man nor book could give an answer to his question, and his mind kept working on it and remained restless until he had devised one himself. This means, of course, that there was no rest for him until the end. In a few cases, however, a satisfactory answer suggested itself, and so a whole system of knowledge was slowly unfolding in him.

His apprenticeship in Verrocchio's studio must have greatly fostered his inquiries in the theory of perspective, the art of light and shade, and the physiology of vision; the preparation of colors and varnishes must have turned his thoughts to chemistry; while the routine of his work woke up naturally enough his interest in anatomy. He could not long be satisfied by the study of the so-called artistic anatomy, which deals only with the exterior muscles. For one thing, the study of the movements of the human figure, which he tried to express in his drawings, raised innumerable questions: how were they possible, what kept the human machine moving and how did it work? . . . It is easy to imagine how he was irresistibly driven step by step to investigate every anatomical and physiological problem. There are in the King's library at Windsor hundreds of drawings of his which prove that he made a thorough analysis of practically all the organs. Indeed, he had dissected quite a number of bodies, including that of a gravid woman, and his minute and comprehensive sketches are the first anatomical drawings worthy of the name. Many of these sketches are devoted to the comparison of human anatomy with the anatomy of animals, the monkey or the horse for instance; or else he will compare similar parts of various animals, say, the eyes or a leg and a wing. Other sketches relate to pathological anatomy: the hardening of the arteries; tuberculous lesions of the lungs; a very searching study of the symptoms of senility.

On the other hand his activity as a practical engineer led him to study, or we might almost say to found, geology: he set to wonder at the various layers of sand and clay which the cutting of a canal did not fail to display; he tried to explain the fossils which he found embedded in the rocks and his explanations were substantially correct. Moreover, he clearly perceived the extreme slowness of most geological transformations, and figured that the alluvial deposits of the river Po were two hundred thousand years old. He well understood the geological action of water and its meteorological cycle.

His work as a sculptor, or as a military engineer (for instance, when he had to supervise the casting of bombards), caused him to study metallurgy, particularly the smelting and casting of bronze, the rolling, drawing, planing, and drilling of iron. On all these subjects he has left elaborate instructions and drawings. He undertook in various parts of northern Italy a vast amount of hydraulic work: digging of canals, for which he devised a whole range of excavating machines and tools; building of sluices; establishment of water wheels and pipes; and his study of hydrodynamics was so continuous that notes referring to it are found in all his manuscripts. He also studied the tides, but did not understand them.

In fact, it is impossible to give even a superficial account of all his scientific and technical investigations, and the reader must forgive me if the magnitude of the subject obliges me to limit myself to a sort of catalogue, for the adequate development of any single point would take many a page. Leonardo's manuscripts contain a great number of architectural drawings, sketches of churches and other buildings, but also more technical matters; he studied the proportion of arches, the construction of bridges and staircases; how to repair fissures in walls; how to lift up and move houses and churches. There is also much of what we would call town planning; the plague of Milan in 1484 likely was his great opportunity in this field; and he thought of various schemes to improve public sanitation and convenience, including a two-level system of streets. Botany repeatedly fixed his attention and we find many notes on the life of plants, the mathematical distribution of leaves on a stem, also beautiful and characteristic drawings of various species. A great deal of the work undertaken for his employers was of course connected with military engineering: hundreds of notes and sketches on all sorts of arms and armor, on all imaginable offensive and defensive appliances; of course, many plans for fortifications and strongholds (how to attack them and how to defend them); portable bridges; mining and countermining; *tanks;* various devices for the use of liquid fire, or of poisoning and asphyxiating fumes. He adds occasional notes on military and naval operations. He had even thought of some kind of submarine apparatus, by means of which ships could be sunk, but the dastardliness of the idea had horrified and stopped him.

No field, however, could offer a fuller scope to his prodigious versatility and ingenuity than the one of practical mechanics. A very intense industrial development had taken place in Tuscany and Lombardy for centuries before Leonardo's birth; the prosperity of their workshops was greater than ever; there was a continuous demand for inventions of all kinds, and no environment was more proper to enhance his mechanical genius.

Leonardo was a born mechanic. He had a deep understanding of the elementary parts of which any machine, however complicated, is made up, and his keen sense of proportions stood him in good stead when he started to build it. He devised machines for almost every purpose which could be thought of in his day. I quote a few examples at random: various types of lathes; machines to shear cloth; automatic file-cutting machines; sprocket wheels and chains for power transmission; machines to saw marble, to raise water, to grind plane and concave mirrors, to dive under water, to lift up, to heat, to light; paddle-wheels to move boats. And mind you, Leonardo was never satisfied with the applications alone, he wanted to understand as thoroughly as possible the principles underlying them. He clearly saw that practice and theory are twin sisters who must develop together, that theory without practice is senseless, and practice without theory hopeless. So it was not enough for him to hit upon a contrivance which answered his purpose; he wanted to know the cause of his success, or, as the case may be, of his failure. That is how we find in his papers the earliest systematic researches on such subjects as the stability of structures, the strength of materials, also on friction, which he tried in various ways to overcome. That is not all: he seems to have grasped the principle of automaticity —that a machine is so much the more efficient, that it is more continuous and more independent of human attention. He had even conceived, in a special case, a judicious saving of human labor, that is, what we now call "scientific management."

His greatest achievement in the field of mechanics,

however, and one which would be sufficient in itself to prove his extraordinary genius, is his exhaustive study of the problem of flying. It is complete, in so far that it would have been impossible to go further at his time, or indeed at any time until the progress of the automobile industry had developed a suitable motor. These investigations which occupied Leonardo throughout his life, were of two kinds. First, a study of the natural flying of birds and bats, and of the structure and function of their wings. He most clearly saw that the bird obtains from the air the recoil and the resistance which is necessary to elevate and carry itself forward. He observed how birds took advantage of the wind and how they used their wings, tails, and heads as propellers, balancers and rudders. In the second place, a mechanical study of various kinds of artificial wings, and of diverse apparatus by means of which a man might move them, using for instance the potential energy of springs, and others which he would employ to equilibrate his machine and steer its course.

It is necessary to insist that most of these drawings and notes of Leonardo's are not idle schemes, vague and easy suggestions such as we find, for instance, in the writings of Roger Bacon; but, on the contrary, very definite and clear ideas which could have been patented, if such a thing as a patent office had already existed! Moreover, a number of these drawings are so elaborate, giving us general views of the whole machine from different directions, and minute sketches of every single piece and of every detail of importance—that it would be easy enough to reconstruct it. In many cases, however, that is not even necessary, since these machines were actually constructed and used, some of them almost to our time.

To visualize better the activity of his mind, let us take at random a few years of his life and watch him at work. We might take, for instance, those years of divine inspiration when he was painting the "Last Supper" in the refectory of Santa Maria delle Grazie, that is, about 1494-1498. Do you suppose that this vast undertaking claimed the whole of his attention?

During these few years we see him act professionally as a pageant master, a decorator, an architect, an hydraulic engineer. His friend, Fra Luca Pacioli, the mathematician, tells us that by 1498 Leonardo "had completed with the greatest care his book on painting and on the movements of the human figure." We also know that before 1499, he had painted the portraits of Cecilia Gallerani and of Lucrezia Crivelli. Besides, his notebooks of that period show that he was interested in a great variety of other subjects, chief among them hydraulics, flying, optics, dynamics, zoology, and the construction of various machines. He was also making a study of his own language, and preparing a sort of Italian dictionary. No wonder that the prior of Santa Maria complained of his slowness!

It so happened that during these four years he did not do much anatomical work, but during almost any other period he would have been carrying on some dissecting. Corpses were always hard to get, and I suppose that when he could get hold of one he made the most of it, working day and night as fast as he could. Then, as a change, he would go out into the fields and gaze at the stars, or at the earthshine which he could see inside the crescent of the moon; or else, if it were daytime, he would pick up fossils or marvel at the regularities of plant structure, or watch chicks breaking their shells. . . . Was it not uncanny? Fortunate was he to be born at a time of relative toleration. If he had appeared a century later, when religious fanaticism had been awakened, be sure this immoderate curiosity would have led him straight to the stake.

But remarkable as Leonardo's universality is, his earnestness and thoroughness are even more so. There is not a bit of dilettantism in him. If a problem has once arrested his attention, he will come back to it year after year. In some cases, we can actually follow his experiments and the hesitations and slow progress of his mind for a period of more than twenty-five years. That is not the least fascinating side of his notes; as he wrote them for his own private use, it is almost as if we heard him think, as if we were admitted to the secret laboratory where his discoveries were slowly maturing. Such an opportunity is unique in the history of science.

Just try to realize what it means: Here we have a man of considerable mother-wit, but unlearned, unsophisticated, who had to take up every question at the very beginning, like a child. Leonardo opened his eyes and looked straight upon the world. There were no books between nature and him; he was untrammelled by learning, prejudice, or convention. He just asked himself questions, made experiments and used his common sense. The world was one to him, and so was science, and so was art. But he did not lose himself in sterile contemplation, or in verbal generalities. He tried to solve patiently each little problem separately. He saw that the only fruitful way of doing that is first to state the problem as clearly as possible, then to isolate it, to make the necessary experiments and to discuss them. Experiment is always at the bottom; mathematics, that is, reason, at the end. In short, the method of inductive philosophy which Francis Bacon was to explain so well a century and a half later, Leonardo actually practiced.

This is, indeed, his greatest contribution: his method. He deeply realized that if we are to know something of this world, we can know it only by patient observation and tireless experiment. His notebooks are just full of experiments and experimental suggestions, "Try this . . . do that . . ." and we find also whole series of experiments, wherein one condition and then another is gradually varied. Now, that may seem of little account, yet it is everything. We can count on our fingers the men who devised real experiments before Leonardo, and these experiments are very few in number and very simple.

But perhaps the best way to show how far he stood on the road to progress, is to consider his attitude in regard to the many superstitions to which even the noblest and most emancipated minds of his day paid homage, and which were to sway Europe for more than two centuries after Leonardo's death. Just remember that in 1484, the Pope Boniface VIII had sown the seed of the witch mania, and that this terrible madness was slowly incubating at the time of which we are speaking. Now, Leonardo's contempt for astrologers and alchemists was most outspoken and unconditional. He met the spiritists of his age, as we do those of today, by simply placing the burden of proof on their shoulders. It is true, for all these matters, his Florentine ancestry stood him in good stead. Petrarca had already shown how Florentine common sense disposed of them; but Petrarca, man of letters, would not have dared to treat the believers in ghosts, the medical quacks, the necromancers, the searchers for gold and for perpetual motion as one bunch of impostors. And that is what Leonardo did repeatedly and most decidedly. Oh! how they must have liked him!

I must insist on this point: it is his ignorance which saved Leonardo. I do not mean to say that he was entirely un-learned, but he was sufficiently unlearned to be untrammelled. However much he may have read in his mature years, I am convinced that the literary studies of his youth were very poor. No teachers had time to mould his mind and to pervert his judgment. The good workman Verrocchio was perhaps his first philosopher, nature herself his real teacher. He was bred upon the experiments of the studio and of real life, not upon the artificialities of a medieval library. He read more, later in life, but even then his readings, I think, were never exhaustive. He was far too original, too impatient. If he began to read, some idea would soon cross his mind, and divert his attention, and the book would be abandoned. Anyhow, at that time his mind was already proof against the scholastic fallacies; he was able, so to say, to filter through his own experience whatever medieval philosophy reached him either in print or by word of mouth.

Neither do I mean to imply that all the schoolmen were dunces. Far from that, not a few were men of amazing genius, but their point of view was never free from prejudice; it was always the theological or legal point of view; they were always like lawyers pleading a cause; they were constitutionally unable to investigate a problem without reservation and without fear. Moreover, they were so cocksure, so dogmatic. Their world was a limited, a closed system; had they not encompassed and exhausted it in their learned encyclopedias? In fact they knew everything except their own ignorance.

Now the fact that Leonardo had been protected against them by his innocence is of course insufficient to account for his genius. Innocence is but a negative quality. Leonardo came to be what he was because he combined in himself a keen and candid intelligence with great technical experience and unusual craftsmanship. That is the very key to the mystery. Maybe if he had been simply a *theoretical* physicist, as were many of the schoolmen (their interest in astronomy and physics was intense), he would not have engaged in so many experiments. But as an engineer, a mechanic, a craftsman, he was experimenting all the while; he could not help it. If he had not experimented on nature, nature would have experimented on him; it was only a choice between offensive and defensive experimenting. Anyhow, whether he chose to take the initiative or not, these experiments were the fountainhead of his genius. To be sure, he had also a genuine interest in science, and the practical problems which he encountered progressively allured him to study it for its own sake, but that took time: once more the craftsman was the father of the scientist.

I would not have the reader believe that everything was wrong and dark in the Middle Ages. This childish view has long been exploded. The most wonderful craftsmanship inspired by noble ideals was the great redeeming feature of that period—unfortunately never applied outside the realm of religion and of beauty. The love of truth did not exalt medieval craftsmen, and it is unlikely that the thought of placing his art at the service of truth ever occurred to any of them.

Now, one does not understand the Renaissance if one fails to see that the revolution—I almost wrote, the miracle—

which happened at that time was essentially the application of this spirit of craftsmanship and experiment to the quest of truth, its sudden extension from the realm of beauty to the realm of science. That is exactly what Leonardo and his fellow investigators did. And there and then modern science was born, but unfortunately Leonardo remained silent, and its prophets came only a century later. . . .

Man has not yet found a better way to be truly original than to go back to nature and to disclose one of her secrets. The Renaissance would not have been a real revolution, if it had been simply a going back to the ancients; it was far more: it was a return to nature. The world, hitherto closed-in and pretty as the garden of a beguinage, suddenly opened into infinity. It gradually occurred to the people—to only very few at first—that the world was not closed and limited, but unlimited, living, forever becoming. The whole perspective of knowledge was upset, and as a natural consequence all moral and social values were transmuted. The humanists had paved the way, for the discovery of the classics had sharpened the critical sense of man, but the revolution itself could only be accomplished by the experimental philosophers. It is clear that the spirit of individuality, which is so often claimed to be the chief characteristic of this movement, is only one aspect of the experimental attitude.

It may seem strange that this technical basis of the Renaissance has been constantly overlooked, but that is simply due to the fact that our historians are literary people, having no interest whatever in craftsmanship. Even in art it is the idea and the ultimate result, not the process and the technique which engross their attention. Many of them look upon any kind of handicraft as something menial. Of course, this narrow view makes it impossible for them to grasp the essential unity of thought and technique, or of science and art. The scope of abstract thinking is very limited; if it be not constantly rejuvenated by contact with nature our mind soon turns in a circle and works in a vacuum. The fundamental vice of the school-men was their inability to avow that, however rich experimental premises may be, their contents are limited;—and there is no magic by means of which it is possible to extract from them more than they contain.

The fact that Leonardo's main contribution is the introduction, not of a system, but rather of a method, a point of view, caused his influence to be restricted to the few people who were not impervious to it. Of course, at almost any period of the past there have been some people—only a very few—who did not need any initiation to understand the experimental point of view, because their souls were naturally oriented in the right way. These men form, so to say, one great intellectual family: Aristotle, Archimedes, Ptolemy, Galen, Roger Bacon, Leonardo, Stevin, Gilbert, Galileo, Huygens, Newton. . . . They hardly need any incentive; they are all right anyhow. However, Leonardo's influence was even more restricted than theirs, because he could never prevail upon himself to publish the results of his experiments and meditations. His notes show that he could occasionally write in a terse language and with a felicity of expression which would be a credit to any writer; but somehow he lacked that particular kind of moral energy which is necessary for a long composition, or he was perhaps inhibited, as so many scientists are, by his exacting ideal of accuracy.

All that we know of Leonardo's scientific activities is patiently dug out of his manuscripts. He was left-handed and wrote left-handedly, that is, in mirror-writing: his writing is like the image of ours in a mirror. It is a clear hand, but the disorder of the text is such that the reading is very painful. Leonardo jumps from one subject to another; the same page may contain remarks on dynamics, on astronomy, an anatomical sketch, and perhaps a draft and calculations for a machine.

The study of Dante is in many ways far simpler. His

scientific lore does not begin to compare with Leonardo's knowledge. The *Divina Commedia* is the sublime apotheosis of the Middle Ages; Leonardo's notebooks are not simply an epitome of the past, but they contain to a large extent the seeds of the future. The world of Dante was the closed medieval world; the world of Leonardo is already the unlimited world of modern man: the immense vision which it opens is not simply one of beauty, of implicit faith, and of corresponding hope; it is a vision of truth, truth in the making. It is perhaps less pleasant, less hopeful; it does not even try to please, nor to give hope; it just tries to show things as they are: it is far more mysterious, and incomparably greater.

I do not mean to say that Dante had not loved truth, but he had loved it like a bashful suitor. Leonardo was like a conquering hero; his was not a passive love, but a devouring passion, an indefatigable and self-denying quest, to which his life and personal happiness were entirely sacrificed. Some litterary people who do not realize what this quest implies, have said that he was selfish. It is true, he took no interest in the petty and hopeless political struggles of his day; Savonarola's revival hardly moved him, and he had no more use for religious charlatanry than for scientific quackery. One would be a poor man, however, who would not recognize at once in Leonardo's aphorisms a genuine religious feeling, that is, a deep sense of brotherhood and unity. His generosity, his spirit of detachment, even his melancholy, are unmistakable signs of true nobility. (He often makes me think of Pascal.) He was very lonely, of course, from his own choice, because he needed time and quietness, but also because, being so utterly different, it is easy to conceive that many did not like him. I find it hard to believe that he was very genial, in spite of what Vasari says. Being surrounded by people whose moral standards were rather low or, if these were higher, who were apt to lose their balance and to become hysterical because of their lack of knowledge,

Leonardo's solitude could but increase, and to protect his equanimity he was obliged to envelop himself in a triple veil of patience, kindness, and irony.

Leonardo's greatest contribution was his method, his attitude; his masterpiece was his life. I have heard people foolishly regret that his insatiable curiosity had diverted him from his work as a painter. In the spiritual sphere it is only quality that matters. If he had painted more and roamed less along untrodden paths, his paintings perhaps would not have taught us more than do those of his Milanese disciples. While, even as they stand now, scarce and partly destroyed, they deliver to us a message which is so uncompromisingly high that even today but few understand it. Let us listen to it; it is worthwhile. This message is as pertinent and as urgent today as it was more than four hundred years ago. And should it not have become more convincing because of all the discoveries which have been made in the meanwhile? Do I dream, or do I actually hear, across these four centuries, Leonardo whisper: "To know is to love. Our first duty is to know. These people who always call me a painter annoy me. Of course, I was a painter, but I was also an engineer, a mechanic. My life was one long struggle with nature, to unravel her secrets and tame her wild forces to the purpose of man. They laughed at me because I was unlettered and slow of speech. Was I? Let me tell you: a literary education is no education. All the classics of the past cannot make men. Experience does, life does. They are rotten with learning and understand nothing. Why do they lie to themselves? How can they keep on living in the shade of knowledge, without coming out in the sun? How can they be satisfied with so little—when there is so much to be known, so much to be admired? . . . They love beauty, so they say—but beauty without truth is nothing but poison. Why do they not interrogate nature? Must we not first understand the laws of nature, and only then the laws and the conventionalities of men? Should we not give more importance to that which is most permanent? The study of nature is the substance of education—the rest is only the ornament. Study it with your brains and with your hands. Do not be afraid to touch her. Those who fear to experiment with their hands will never know anything. We must all be craftsmen of some kind. Honest craftsmanship is the hope of the world. . . ."

Conditions for Discovery" is based on an address given by Dr. Philip H. Abelson, Director of Carnegie Institution's Geophysical Laboratory, on the occasion of the dedication of the American Medical Association's Institute for Biomedical Research, in Chicago, on October 11, 1965. The address was published in the Journal of the American Medical Association *for December 27 of that year, and is republished here in somewhat shortened form.*

Dr. Abelson was born in Tacoma, Washington, in 1913. After earning a bachelor's degree in chemistry and a master's degree in physics at the State College of Washington, he went to the University of California at Berkeley to study at Ernest O. Lawrence's Radiation Laboratory, with its famous cyclotron. While working toward his doctorate in nuclear physics at the Laboratory, he made the first American identification of the products of uranium fission.

Dr. Abelson joined the Carnegie Institution in 1939 and helped design and build a cyclotron for its Department of Terrestrial Magnetism. In 1940 he and Edwin M. McMillan discovered the element neptunium. During World War II Abelson worked for the Naval Research Laboratory, where he developed the liquid thermal diffusion method for the separation of uranium. This was one of the methods adopted by the Manhattan Project to produce enough enriched uranium for the atomic bomb. At the end of the war he prepared blueprints and a feasibility study for an atomic submarine. This Abelson Report spurred Admiral Hyman Rickover in his crusade for a nuclear submarine fleet. For his contributions to the war effort the Navy in 1945 gave Abelson its top civilian award, the Distinguished Civilian Service Medal.

Returning to the Carnegie Institution in 1946, Dr. Abelson became chairman of a biophysics group at the Department of Terrestrial Magnetism. With his colleagues there he used radioactive isotopes to study how cells select chemicals from their environment and organize them into new cells. In 1953 he turned to yet another field when he was appointed Director of Carnegie's Geophysical Laboratory. Since then he has been particularly interested in biochemical evolution and the origin, antiquity and history of life on the earth.

Besides his research and administrative responsibilities at Carnegie, Abelson edits Science, *the magazine of the American Association for the Advancement of Science.*

Dr. Abelson is represented by two essays in this volume.

26

Philip H. Abelson

CONDITIONS FOR DISCOVERY

From the *Journal of the American Medical Association*, December 27, 1965.

. . . An understanding of some of the factors which affect creativity is vital to the success of any research group. Creative insight is attainable only by individuals and so most discussions of creativity focus on the individual and his mental processes. But in the realities of today's world, scientists who are creative rarely dwell alone in isolated ivory towers. All to some degree receive stimulus, information, and help in making judgments from others in their immediate environment. I believe that these interactions are highly important and, with the evolution of science, they are becoming even more so. . . .

The usual discussion of creativity describes it as occurring in four steps:

1. *Preparation,* involving thorough investigation of the problem by reading and experiment.

2. *Incubation,* involving a conscious and unconscious mental digestion and assimilation of all pertinent information acquired.

3. *Illumination,* involving the appearance of the creative idea, the creative flash.

4. *Verification,* involving experimental testing of the creative idea.

In discussions of creativity most authors emphasize what is often the spectacular phase, illumination, perhaps because people are fascinated by the spectacular.

27

Major creative flashes have been noted by many great scientists; undoubtedly they constitute an important phenomenon. Nevertheless, dramatic illumination, though representing a culmination, is for most scientists only one part, perhaps even a minor part, of creativity. The other steps are also essential. Without preparation, and incubation, there can be little illumination.

One crucial element in creativity is judgment. When an experimental scientist seeks to be creative, he must make and implement a series of judgments. The complexity of his needs can be seen by following some of the steps. First, he must decide upon an area to investigate. Many scientists find that they can name dozens of possibilities. It is necessary somehow to focus on one or only a few areas to study at one time. Obviously, the choice is crucial, for diligence that may be expended fruitlessly in one area might be richly rewarded in another. Having selected the major topic, the investigator must decide what to read, and how intensively. Some research scientists get lost in the literature. They cannot make proper judgments about what to read, what to skip, what to believe, what to be skeptical about. Next, the scientist must decide what approach to take. Again, judgment must be applied, and even some creativity, for a good method is often the key to major discovery.

Once he begins to experiment, the investigator faces more judgments. Are his methods sound, or are there hidden defects? Chance observations may suggest byways. Should he pursue them? Sometimes results that initially seem valid are not repeatable. Should he build a new apparatus to pursue his original goal, or should he turn to another problem? The creative scientist is daily faced with major and minor choices. Either he makes good choices as a result of deep thought and the use of every available aid of consultation, or he makes bad choices by default.

Good judgment is rarely on tap. It requires a weighing

of obvious alternatives and even the invention of new alternatives. Usually the best judgments come after extended wrestling with a problem and in an environment free from distractions. In today's world, judgment and reflection are often crowded out by activities that represent more fun for the scientist—more experiments, administrative matters, committee work. In the face of the sharp competition of distractions, few scientists exhibit as high a quality of judgment as they are potentially capable of. To exert good judgment requires self-discipline which, in turn, rests on adequate motivation.

Motivation is essential to creativity; without it, even the best minds accomplish little. With adequate motivation comes the self-control necessary to tap inner resources. Creative effort differs from most other activities in requiring unusual discipline. People in other walks of life may go for long periods without exercising much self-control, but a creative scientist must take himself in hand. The initiative must be his. No foreman can tell him how to think or what to do next. He has to do his own thinking and make his own judgments. If he fails to exercise proper self-discipline, the deficiency may not be obvious to others. He may appear for work at the usual hour, go through all the accustomed routines, attend seminars, read the literature, and give the appearance of creative effort. But if his mind is elsewhere, his activity may be only a facade for inertia.

Related to the need for self-discipline are qualities of patience, courage, and willingness to take the punishment of disappointment. In the present era of science there is pressure to build bibliographies. The easy way to do it is to carry on research that is merely a small extension of what is already known. Then the scientist does not have to think very deeply; he can let a technician perform the work; and at the same time he feels some security as a contributor to science. The path of courage lies in choosing a difficult but fundamental problem and working at it even though the walls of confusion seem insurmountable. The person who undertakes such a task must be capable of living with disappointment. He must be able to cope with the unhappiness that follows failure of what seemed to be promising approaches. Even after an extended period of work from which there is no obvious accomplishment he must be capable of summoning the necessary stamina to persist. The inner resources on which the creative person calls in order to continue after repeated failure can only be tapped by deep motivation.

How are people motivated? To that question there is no single easy answer. First of all, the great variation in the intrinsic capacity of individuals to respond to motivating stimuli must be recognized. Little can be done with a man devoid of character. The capacity to be motivated has its roots in genetics and is nurtured by environment.

One of the most effective agents is mutual stimulation arising from the interaction of two or more individuals. Given the right circumstances, human beings can interact in extremely constructive, helpful ways. The stimulus can be provided by another individual, by a small group, by a larger group, or by the general community. The interactions can produce enhanced motivation; they also can provide other factors (judgment, for example) that are essential to the creative scientist. . . .

For years I have been impressed by the enormous scientific contributions of three Hungarians—John von Neumann, Eugene Wigner, and Leo Szilard. Since all came from Budapest, I asked Professor Wigner about possible interaction among them. He told me that there had been a strong mutual stimulus. Wigner and von Neumann had known each other during high school days in Budapest. Later they had both gone to college in Berlin. Being foreigners, and not feeling part of the social structure, they became especially close. In fact they roomed together. Professor Wigner told me that he learned more from von Neumann than from any other man, and though

Wigner would be modest in his estimates, I am sure that von Neumann learned a great deal from Wigner. Later these two became associated with the highly imaginative and stimulating Leo Szilard, and the three were very close. They found excitement in ideas and mutual stimulus in arguments and discussions with one another. Finding their surroundings uncongenial, they discovered in creative science a perfect vehicle for escape from the rest of the world and, at the same time, a wholly satisfying common ground for excitement.

Dr. Wigner, in discussing the matter of creativity with me, emphasized this matter of discontent, of the importance of not quite fitting into the environment. He also suggested that perhaps the success of the two Chinese Nobel-prize winners, Lee and Yang, resulted from a similar relationship, and Professor Yang has confirmed the surmise. Lee and Yang had known each other in China during the forties. They reestablished contact at Chicago in the late forties. There was an intense interaction, which was later noted at Princeton. It was highly constructive and contributed much to their obtaining the Nobel prize.

Another famous example of constructive interaction involved Enrico Fermi and a group of physicists in Rome during the late twenties and thirties. Fermi first began his work in Rome in 1926 shortly after obtaining his Ph.D. degree at Pisa. In the preceding generation there had been no first-class physicists in Italy. Yet, in a few years there gathered around Fermi men who were to gain renown. Fermi himself, moreover, was to become one of the greatest physicists of all time. When the group began to interact, there were only a few indications of future greatness. Fermi had already shown signs of being a more than ordinary man. His first associate, Franco Rasseti, even in graduate school showed a touch of genius. However, Emilio Segré, who was later to receive the Nobel prize, was still an undergraduate, as was Edoardo Amaldi, now one of Europe's leading physicists. Fermi, of course, was the leader of the group, but there is little doubt that he got stimulus from the others and that they in turn benefited from association with him. The group was much stronger than the sum of its parts. Moreover, the stimulus of working together for a time had a permanent beneficial effect on all the members—an effect that lasted long after the vicissitudes of politics and war broke up the association.

An experience of group interstimulation was very valuable to me personally. Shortly after World War II, three other nuclear physicists and I joined with a biologist to form a biophysics group at the Department of Terrestrial Magnetism of the Carnegie Institution of Washington. The physicists knew no biology and almost no chemistry, and it was ridiculous to attempt biophysics research in a department devoted primarily to geophysics. Professional biologists and biochemists who learned of our activity felt free to comment caustically; one distinguished biochemist was heard to describe us as "a group of wistful physicists out looking for the secret of life." At the same time our former colleagues in nuclear physics shook their heads sadly at our insanity in abandoning an established professional position. The semihostility of the outside world provided an incentive to knit the group together. Naturally, we were under all kinds of pressure to establish a niche for ourselves and to win the respect of a new body of professional colleagues. Embarking on a course of self-education, we taught each other physical chemistry, biochemistry, and microbiology. Our conversation and our thoughts were centered on the experimental work we were doing and on its interpretation. While each usually worked on his own individual projects, all helped one another in many ways—with suggestions, enthusiasm, and judgment. We gave mutual support by fostering the concept that all of us were doing important, exciting, and significant work. We felt lucky to be having so much fun while being paid for it. Soon we were pioneering in studies of biosynthesis in microorganisms, and in a few years we were making worth·

while contributions. Our book entitled *Studies of Biosynthesis in Escherichia coli* came to be regarded as a bible for students of microbiology and received favorable comment from all over the world.

These examples of mutual stimulus could be multiplied. Most of us are responsive to our environment, and group stimulus seems to be one of the best ways of enhancing creativity. In the examples I have cited there are several components that I believe to be essential. First, there is a banding together of individuals to create a microenvironment. Those within the group can get satisfaction and a meeting of the need for human fellowship from within this tiny intellectual island. It is then possible to adopt the attitude that the value system of the rest of the world does not really count, that a common interest in advancing an area of science is the most important and the only tenable activity in which the individual can engage. Moreover, if a member temporarily loses his fire, he can soon regain it, for at all times there are some present who are brimming with enthusiasm.

Judgment can usually be sharpened through group interactions. Related to this is the matter of "blind spots." A man may think deeply about a problem but fail to cope with it successfully because of lack of some small crucial idea—perhaps an item of knowledge, perhaps simply a slight failure in analysis. Discussion of the problem with others can provide the single item that opens new vistas of comprehension.

Bringing a group together does not invariably produce this favorable mutual stimulus. Men who have the capacity to create must have their share of pride and egotism. In a closely knit, tight environment, tensions and rivalries are always latent. Indeed, the greatest hazard in group activity is internal dissension. When adrenalin begins to flow, creativity goes out the window. A person engaged in a serious personal clash of wills invariably gives first priority in his thinking to the matter of issue. If you are in a fight, you get ready to fight. All too often the conflict escalates to implacable hatred, to the detriment of the individuals directly involved and all those around them. The carping critic, the jealous troublemaker, usually destroys far more than he creates, and his influence can wither the creativity of many around him. If you have a troublemaker in your midst, throw him out.

The tendency toward dissension can be countered in various ways that basically amount to the same thing. It is necessary that the individuals become so desirous of accomplishing common goals that they suppress their natural egotism, or, alternatively, each must receive so many benefits from an association that he believes self-interest to require a smooth relationship.

One clear-cut example of a common goal is seen in wartime, when national energies are concentrated to meet a common foe. Those who worked in the great wartime laboratories remember that the activities were relatively free of the usual frictions. In some of the examples of interaction previously cited, there was a common element of group unity in the face of an unfriendly or drab environment. Rivalry between institutions can be another healthy way of promoting internal accord and purpose.

An effective means of ensuring peace within a small group lies in the choice of its members. They should be people whose background and temperament are different. One, for instance, may be a skilled experimenter, and another a great enthusiast. A third may have excellent judgment, the ability to recognize which leads are likely to prove valuable and to quietly dispose of the trivial, unsound, or sterile. It is usually desirable that the group contain a compromiser or peacemaker as well.

It is also desirable that the various members of the group have somewhat differing bodies of knowledge so that they can teach one another. By assembling people of sufficiently different competence, almost encyclopedic knowledge can be provided in even a small group. Individuals having different talent and knowledge are intrinsically less competitive, for each has the security of knowing that he has a special contribution to make.

Of all the desirable components in a research group, the most crucial is that of the enthusiast. In some groups, such as the one headed by Fermi, the leader is the enthusiast. But that is not essential. The spark plug can be almost anyone in the group. Recently a chairman of a department in a great university told me of an observation that he and his staff had made. They had become aware that the current crop of graduate students was performing far beyond anything seen in the last 20 years. After some searching self-analysis, they were forced to conclude that the inspiration was coming not from the faculty but from a particular graduate student. The young man was not much as a student. Indeed they had considered flunking him out. But he was getting research done and he was firing up all the other students to do likewise. . . .

Constructive group interaction is important, but it is not the only essential component in achieving and maintaining a creative situation. One need is for continuing education among mature scientists. At a university, scholars are exposed to constant stimulus, and if they are willing to exert themselves, they can remain broadly aware of important developments in science. Members of this Institute must meet the problem of continuing education and provide mechanisms for it. Otherwise they face obsolescence. The development of effective means of retooling and sharpening the mature scientist is worth a good deal of thought and effort.

Highly intelligent, highly motivated individuals with natural good judgment can probably attain creative insights in almost any environment. Most people, however, are much influenced by surroundings and the circumstances under which they work. In any research effort there are usually periods of relatively routine work, and often they are essential. The crucial effort, however, is that which goes into the judgment of interpretation, the decisions of what to do next, especially the decision to embark on innovation. Such decisions have their highest quality when they evolve from intuitive insights. Every man I have known who is markedly creative has devoted a substantial part of his time to thinking about his work. I believe that complete immersion in a problem, at least periodically, is essential to developing a person's best potential.

In our efforts at organizing research we probably make our greatest mistakes in failing to provide for the need for total immersion in creative thought. In choosing group leaders, section heads, and others higher up on the administrative ladder we select men who have demonstrated ability as creative scientists. We then saddle them with the need to make a continuing series of major and minor decisions, and we provide them with an ever-ringing telephone and a chattering distraction for a secretary. In the process we destroy most of their potential for creativity. In fact, one of the greatest things that scientific laboratories could do would be to abolish the telephone for most of the day.

During the past 11 years as director of a laboratory, I have been determined to carry on personal research no matter what the administrative commitments. I have learned that I work most effectively if I concentrate on doing one type of job at a time, giving it an overriding priority. When I am working at research, I turn my back on administration and try to devote at least a solid week to the effort. During that period I often do not even open my mail—much less respond to it. If I answer the phone, which usually I don't, I do not engage in long conversations. During such periods I find it particularly useful to do technical work in the evening or to read just before going to bed. Very often I am rewarded in the morning by new ideas.

From these observations I would suggest that research administration be organized so that all creative individuals have frequent opportunities to attain total immersion in technical problems. Administrative responsibilities should be attended to periodically, but with free spaces in between.

Another personal observation is a seasonal variation in creative capacities. Almost every good idea I have had, fresh approach I have initiated, during 30 years, has occurred in January, February, or March. Knowing this, I try to avoid making interfering commitments for that period but devote as much time as possible then to laboratory work.

Many scientists find it useful to have two or three deadlines to meet each year. At our laboratory we have an annual report that we take very seriously. The report, a technical discussion of all work that has made substantial progress during the year, is sent to more than 2,000 persons and establishments throughout the world. When the deadline for it approaches, our staff puts on a tremendous spurt of effort to complete studies for inclusion in the report. Another type of deadline occurs as the staff prepares papers for technical meetings, when there is another burst of effort, and men turn out several times more work than usual. . . .

Another important factor in creativity is the achieving of balance among the various components that go into the creative act. Some men spend so much time reading the literature that they never get around to doing anything. Others live a life characterized by drive, drive, drive. They are so busy that they have no time to consider what is worth doing or what is significant about what they have done.

The need is not only for a balance among the various activities, but for an appropriate sense of timing of them. An ancient biblical philosopher stated the matter very well in the third chapter of Ecclesiastes (3:1-2):

> To everything there is a season and a time to every purpose under the heaven: a time to be born and a time to die; a time to plant, and a time to pluck up that which is planted.

These ancient words are particularly relevant in scientific research. There is a time for reading the literature, a time for writing it. There is a time for reflection, a time for deadlines; a time for broad awareness, and a time for single-minded action.

Too many scientists perform these functions in a piecemeal, jumbled, hit-or-miss fashion. Most individuals could markedly improve their effectiveness by carefully planning their activities so that they successively immerse themselves in the various essential functions of creativity. . . .

D*r. Merle A. Tuve is one of our time's outstanding explorers of the secrets of earth, atmosphere, radiation, space, and matter.*

He was born June 27, 1901, at Canton, South Dakota. His interest in science led him to enroll as an engineering student at the University of Minnesota. He did graduate work in physics at Princeton and Johns Hopkins, and received his doctorate from Johns Hopkins in 1926. He has received honorary degrees from a number of institutions, but perhaps the most unusual was awarded by Carleton College, Minnesota, in 1961 when, with his sister, Rosemond Tuve, and two brothers, George Lewis Tuve and Richard Tuve, he received a special family award for excellence in scholarship, as well as a Doctor of Laws degree.

In the summer of 1925, while a 24-year-old instructor of physics at the Johns Hopkins University at Baltimore, Maryland, Dr. Tuve participated in an experiment that was to have significant results in providing further understanding of electrical layers in the atmosphere, and which led, among other things, to the development of radar. See "Radio Echoes" elsewhere in this volume.

He joined the Carnegie Institution in 1926 as a member of the staff of the Department of Terrestrial Magnetism. He turned his attention to the production of high-energy particles, helped develop the first direct-current high-voltage "atom smasher," and went on to measure the forces inside the atomic nucleus.

He took leave from Carnegie Institution from 1940 until 1945 to serve as a section chairman of the Office of Scientific Research and Development. He managed the development of the proximity fuze — a radio transmitting and receiving set fitted into an anti-aircraft shell and rigged to detonate the shell as it approaches the target. In recognition of his war work Tuve was awarded the Presidential Medal of Merit.

He returned to the Carnegie Institution in 1946 as Director of the Department of Terrestrial Magnetism, a position he held until his retirement in 1966, when he became the second Distinguished Service Member of the Carnegie Institution. Since 1946 his research has included radio astronomy and the design of image tubes for telescopes, and various aspects of geophysics and biophysics.

Merle A. Tuve

PHYSICS AND THE HUMANITIES—
THE VERIFICATION OF
COMPLEMENTARITY

ks on receiving the Third Cosmos Club Award,
ngton, D. C., May 9, 1966.

of
the
ven
any
but
com-
llable
ividual

they
ultimate
the social
mpromise
rd higher
rience. We
lems of the
so universal
tory ways in
ns, are com-
r us to resolve
ge concerning
particles that
accelerators of

e survey the actual life-matrix in which our think-
d by which it is obviously conditioned, we find
lly that life for each of us is much wider than
ofessional interest and competence. In addi-
sometimes humdrum, and sometimes ec-
g which fills most of our waking hours,
much involved with specialized areas of
hich are classified as belonging to other
nterest or professional competence.
a great variety of experiences during
have found myself concerned with
ized areas quite outside of my own
se topics there have been several
s appropriate for my talk here
. . . questions of general public
y what right of thoughfulness
r myself qualified to discuss
ny of experts which is the
hat I should speak out as
ator. . . . It has seemed

best . . ., however, for me to examine the ways in which physics makes its most basic contributions to modern society, contributions to the universal search by every man, when he is not totally preoccupied by hunger or by some immediate threat of catastrophe, for fullness and meaning in his life, the search for richness of experience, set in categories of value—in short, the search for significance in his life.

Most of us, at first encounter with questions about the contributions of physics, think at once of atomic power or atomic bombs ("to preserve the peace"); then we think of the marvels of electronic communication, then of air transportation and space travel, then perhaps of computers and the automation of industry. Then the mood darkens and we think of the troubles of human displacement and adjustment, and the accelerating pace of life which dominates and threatens us, as the world's most favored few, and we wonder if the gift of modern science yields to human beings anything power and the impersonal domination of more and more plex technological systems, seemingly with uncontr destinies of their own, as mankind cowers, and the ind feels more and more lost and estranged.

Oddly enough, these are not problems of scien are all *problems for the humanities* to resolve to their roots in human hopes and needs, and then to guide sciences and governments and peoples toward co and resolution, and finally, to lift men's eyes tow levels of brotherhood and richness of personal exp all know this, and we all recognize that these pro innate plus the cultural characteristics of mankind in some ways, and yet expressed in such contradi the various groupings of cultures and of natic mandingly urgent. They are much more urgent f constructively than our needs for more knowle the extremely rare and fantastically short-live we knock out of atomic nuclei using the hug high-energy physics, or our needs for the satisfaction of our competitive drives and the possible answering of a very few questions of very specialized nature by our present vast Federal efforts for, and commitments to, of all startling things, the man on the moon. This particular error of public emphasis seems absurd, on the face of it, when we consider the enormously destructive potential of the emphatic questions of human maladjustment and disagreement now conspicuous throughout the world, and the new dimensions of social power and individual weakness. Yet, so to speak, we fiddle while Rome burns. This kind of distortion in our public and private lives is one example of the ways in which technology, unless guided by philosophy, can threaten and defeat the good life.

I am known to be one of the early and steadfast critics inside the general fraternity of the physical sciences—in "big science" circles, if you favor that epithet—opposing the general notion that just because we think a new big thing can be done we must now do it, and, correspondingly, our government must pay the bill. I think that in science, as in all other human affairs, a touch of austerity is a necessity for health, and our powers for discrimination and relevance must be fully exercised, or we will surely spend our efforts on conspicuous or gaudy projects, not on those of basic concern.

So I sat in my study in the quiet of the night and asked myself these questions: What are the really basic contributions of modern physics to the areas of the good life, to the full and significant experience of living? What specific elements of importance does physics yield for individual persons, for you and me and for whole societies, among them our own Western cultural groups? Has physics made contributions which may also be relevant and important to those individuals who have inherited the magnificently different traditions of the Far East, and view life from a background of Oriental philosophy? Has it value and meaning even for the fragmented groups now seeking to leap from narrow tribal confines to a full participation in world society? What do I really think are the outstanding bequests from physics to the immediate future of human satisfaction and fulfillment? Are any of these contributions possessed of such sure inherent qualities of permanence that they will continue to contribute to the lives of men, and to their sense of wholeness and unity with Nature and with God, over the uncertain long and upward reaches of the future?

This whole area of personal contemplation and assessment—indeed the entire development of the mental and emotional content of significant living—the ways by which we enjoy and enlarge the richness of human experience—we seem to have allocated to the general area referred to as the humanities and the fine arts. As I sat there in my study, my thinking slowly came to a focus on one very important contribution of modern physics in this intimate field of personal awareness and significance. This contribution is called *complementarity*.

As a physicist I am an experimentalist, but as such I have always been much concerned, of course, with interpretation and theory. Theoretical physics has a language of its own, almost impenetrable, but, like the doorkeeper at the house of the Lord, I have long stood at the door of theoretical physics with my ears open.

There is no necessity for me to paint mental pictures showing the vast scope of experimental physics, ranging from distant quasi-stellar objects of astrophysics to the mesons and strange particles of nuclei and high energy physics, and even farther to the deceptive simplicities of the DNA and RNA of biophysics. Instead, I shall simply draw your attention to one basic clearcut result of modern theoretical physics which has, in my judgment, immeasurably wide and deep significance for every man.

The verified necessity for us to accept two very different views of natural events, mutually irreducible one to the other, a necessity which has been given the name of *comple-*

mentarity, is one great gift of physics in our epoch to the thinking of all humanity. In what follows I shall endeavor to indicate something of the content of this sober remark.

I can express tonight only feebly what has been much better said by others, namely, that the ultimate effect on human life and endeavor and satisfaction of this precious fragment of new understanding, based irrevocably on simple experiments of modern physics, may be far more profound than any of the technological offshoots of atomic energy or electronics or automation which also trace their genealogy to the curiosity and the research studies of the physicists. A number of my friends have expounded the nature of belief or conviction, the vast range of our awareness and the limited fraction of it which is covered by physics, and have gone on to expound the pertinence of the concepts of complementarity and the uncertainty principle in relation to man's own personal acceptance of the dichotomies or antinomies of human awareness. Among these friends are Warren Weaver and Robert Oppenheimer, to whom I owe, without their knowing it, much of the courage needed to make this address.

Over the years I have been privileged to know personally many or most of the giants of this splendid modern enterprise of human understanding, the demonstration of the intrinsic fact of complementarity in the physical world. They were few in number. Perhaps they were fewer than a dozen, all associated intimately and frequently with Niels Bohr, of Copenhagen, who made many visits to the U.S.A. and shared himself with all of us. I personally was privileged to share only in a very small way in the yeast of these developments as they happened, mostly as a listener, although later my colleagues and I shared the related pleasure of making, here in Chevy Chase, the first quantitative measurements of proton-proton and proton-neutron force interactions—the first measurements of the binding force which, by $E = mc^2$, makes atomic nuclei weigh less than the sum of their constituents. (Incidentally,

I might remark that I then proposed in a letter to *The Physical Review* that because of this weight loss the nuclear forces might be called the force of levity, which is surely as fundamental to all Nature as the force of gravity! To this day, however, my sober-minded scientific friends have elected to disdain that gay remark, much as though I had broken out with applause in church.)

The studies of nuclear physics using high-energy particles from accelerators really came after the great ferment of struggle and insight brought about in atomic physics by the nuclear atomic model of Rutherford, the Bohr atom model, and the mathematical formulations of Bohr, Pauli, Heisenberg, Schrödinger, deBroglie, Dirac, and a few others in the period 1911 to 1930. The most intense period was perhaps 1924 to 1928. By these men an abrupt and permanent mutation in human thinking was completed about thirty years ago, and was expressed in two or three basic ideas or statements referred to as the principles of correspondence, of complementarity and of indeterminance.

I shall indicate only very briefly the nature of these ideas. The Bohr atom of 1912, with its quantum transitions between steady states, quite irrational and absurd from the viewpoint of classical Newtonian mechanics and Maxwellian electrodynamics, proved experimentally and empirically correct beyond all possibility of doubting, yet defied all possibility of understanding in terms of previous dynamics. Various mathematical formalisms were devised which simply "described" atomic states and transitions, but the same arbitrary avoidance of detailed processes, for example, descriptions of the actual *process* of transition, were inherent in all these formulations.

During this same period, around 1925, it was demonstrated by G. P. Thomson in England and by Davisson and Germer at the Bell Laboratories in New York that electrons undeniably behaved as waves in certain types of experiments which asked typical questions formulated in terms of wave ideas, such as the positions of interference maxima and minima, yet the arrival of the electrons at these wave positions was never possible to show as waves but always, and only, as separate and discrete particles. After some years of puzzling as to the nature of this wave description, it was realized that the most nearly correct statement was that they were waves of probability. The wave does not predict the actual appearance of an electron as a particle, but only the relative probability, given large numbers of electrons, for the appearance of electrons at the positions indicated by the intensity of the computed waves for these positions. . . . This *complementarity* of the electron both as a wave and a particle is a basic, inescapable dichotomy, on a par with the irrational dynamics of Bohr's nonradiating stationary atomic states and the lack of any conceptual notion which might describe the actual process of the transition of the Bohr atom from one stationary state to another.

Other detailed analyses, especially by conceptual experiments (in contrast to laboratory experiments), led to the inescapable realization that the process of observing or measuring, when we deal with space and mass and energy of atomic dimensions, affects the result of the measurement itself, much as wave questions give wave answers and particle questions give particle answers, when we examine the behavior of electrons. This effect of the observer was most graphically expressed by saying that if we seek to know the exact position of an atomic event, such as a collision, we must be content to know nothing about the momentums involved, or if we ask in detail about momentum exchange, we must be content to remain ignorant of exactly where the event took place. This kind of technical statement is one expression of the basic fact that we cannot learn or predict atomic processes in full detail, which is Heisenberg's Principle of *Indeterminance*.

For ordinary massive bodies the old Newtonian dynamics and Maxwellian electrodynamics were not altered by any of these discoveries or conceptual experiments, even out to extremes of large size and mass and energy. This recognition was expressed by the word *correspondence,* which affirmed the fact that all atomic equations must reduce to the classical equations (as earlier revised in certain ways by Einstein and by Planck) when aggregates of matter, large compared to single atoms, are considered.

By 1931 this radical change in our basic philosophical approach to the laws of Nature was complete.

Now for 35 years man has slowly begun to recognize the implications of this discovery that there are immutable and inescapable dichotomies in his views of the same objects in the physical world, depending on the kinds of observations he makes, the kinds of questions he asks.

The shockingly new and far-reaching discovery, for one's everyday philosophy of living, is that the most abstruse calculations and knowledge of the most detailed and unexpected phenomena of the atomic world which we observe and measure have led unequivocally to a profound verification of an age-old inner awareness which is the most immediate and commonplace philosophical experience of every individual person.

From the time we were adolescents we have all been troubled by the obvious fact that there are different modes of examination of our individual experiences in living, alternate modes roughly characterized by the differences between the sciences and the humanities, but on an internal basis, which we experience as the contrast of *rational* factors with *emotional* factors. And now we know, from the most detailed and profound examination of the outside, objective, physical world around us, that the true answers about reality are intrinsically determined by the way we frame the questions. Furthermore, we find that these answers can even be *logically incompatible* with each other in very fundamental ways, characterized by the nature of the observations or the assessments we make, which are chosen to fit the nature of the particular questions to which we address our attention.

When we ask whether electrons are waves or particles, or whether ordinary light is emitted and travels as waves or as particles, we learn from both theory and experiment that we are compelled by Nature to say, with clear inner contradiction, *both,* but *not* both at once, only one or the other, depending on the *question* you ask—and this you *must ask first.* [This is like the definition of infinity: you name a number first, then I can name a larger number. This describes a process, not a finite stationary state which can be localized and identified.] We have the same experiences with categories of our own personal thought or attention; blood may rush to our cheeks and our hearts may pound with embarrassment, but this is not equivalent to, or interchangeable in any way, with our intense awareness of the moment or our thoughts about the faux pas which precipitated the blush. The motivation, and the initiative for retrieving the situation, again are not in the same category with the biochemical processes which activate our muscles as we retreat from the encounter, however simultaneous in real time these parallel realities may occur or exist.

In biophysics and biochemistry we have been pressing forward in the search for the "physical basis" of various self-propagating or self-activating expressions of living matter, all the way from genetics and the biological coding of DNA and RNA for the synthesis of proteins and the determination of enzymatic behavior, to the roles of RNA and protein synthesis in the formation and retention of memory, which is the acquisition of knowledge by an individual. Half of the research effort of our own Carnegie Laboratory in Chevy Chase for the past 20 years has been devoted to these basic aspects of biophysics. Yet, no man can claim that a description of the physical processes which are simultaneous with a thought or an emotion are fully and identically the *same* as that thought or emotion.

This is not the kind of resolution a physical scientist can seek for the mysteries of life and thought and feeling. Far more acceptable is the intuitive answer men have always given, from the most primitive savage to the most sophisticated intellectual, that things of the mind and spirit of man have a reality all their own, parallel with, but not the same as, our participation in the physical world. Here again, the answers you find are basically determined by the questions you ask. This is an age-old expression of a similar kind of *complementarity,* sensed and expressed and never "resolved." Modern physics, unexpectedly, and in the experimental world of the laboratory, has provided an unequivocal demonstration of the finite or limited nature of our human possibilities for "understanding" the world in which, unasked and without choice, we find ourselves. The message is clear and direct, and leads both to confidence, to faith, and to humility. There *are* different ways, conflicting ways too, in their very essence, by which we view the world around and in us, but the true answers will be framed in terms which fit the questions which we ask.

This verification of complementarity, however, is decidedly not a license to think whatever you please, or to believe to be true whatever ideas may appeal to you. Atomic physics has given an unequivocal demonstration that two sets of ideas, as contradictory as diffusely spread waves and particles concentrated at singular points, are both true descriptions of the material aspect of Nature, complementary views of reality, each of which is necessary in its own turn to answer special questions men have learned to ask. But this does not mean that reality or truth is so unlimited in scope, so multiple-valued or various in its facets, that every fancied description is as valid as any other. Instead, it comprises an unexpected but very specific proof of the limits of the area of authoritative competence of those individuals whose excessive confidence in the rational or scientific approach to the mysteries of Nature and comprehension leads them to discard all the other aspects of awareness and life experience. At the same time, it constitutes an invitation and a challenge to esthetics and the humanities to seek new and different expressions of the spirit of man. It is a fresh encouragement to those among us who are continuing the long search for criteria of validity in those vast realms of ideas and personal awareness which cannot be reduced to axioms and logic and measurement, which defy classification systems, which often elude us and largely disappear when we subject them to the rigidities of language.

Truth and beauty in science and technology comprise only a minor fraction of the total of truth and beauty in human awareness. It is a major step, of course, and vastly comforting, for physics to reach a reasonably comprehensive description of the small and large examples of material things which surround us in this physical universe, and to have these descriptions substantiated by more and more observations. Complementarity and correspondence principles indicate the finite limitations of human concepts, and indeterminance expresses an appropriately corresponding humility. It has taken about 2,500 years of interest and puzzlement and thinking to reach this tentatively satisfactory, though highly circumscribed and limited, world view.

There is something quite poetic about the fact that the capstone of all this achievement of ideas in the realm of the rational aspects of our thought-processes, viewing primarily those material items around us which are not endowed with the special property we call living or life, is this demonstration of the finite or limited scope of our idea processes. We have wondered in physics about the kinds of research which will be interesting, or even possible, 50 or 100 years hence. Most of the basic laws seem known, and the simple things, the one-man accomplishments, seem to have been already studied and done.

This state of affairs is in great contrast, I would guess, to the situation in the humanities and esthetics and the fine arts. Perhaps there are the equivalents of Newton's Laws and

Galileo's telescope and the conservation of energy in these other areas of human contemplation, but if so they seem pretty much obscured by the overlaid accretions of many overzealous generations in each of many different cultures. The whole concept of such equivalents I would myself expect to be erroneous. Surely we have, however, some key concepts for organizing our thoughts and feelings in these areas, concepts such as justice, love, freedom, beauty, honor, and their opposites, and we have systems of thinking, full of similarities and agreements, as outlined in various religions and philosophies, which most certainly are not all wrong or pointless. In fact, the generally accepted public philosophy largely governs the actions of a people, and difficulties have arisen among men and societies whenever the mechanisms of society became too visibly different from the ideas held by that society in the realm of the humanities and personal convictions.

The chief point to be made, perhaps, is that if physics is running out of interesting problems, the situation is quite the opposite in these other areas of personal awareness and ultimate concern. A subsidiary point, which is the theme of this talk, is that physics has put forward, as a permanent major feature of man's knowledge of matter and space, a special disclaimer, a special denial of the uniqueness and all-encompassing validity of the rational and material ideas of science, vaunted by some philosophers and helplessly accepted by so many others less well equipped to judge. Modern physics has given to all a specific demonstration that more than one set of ideas must be used if our finite minds are to view and comprehend in some increasing measure of completeness the fabulous complexity and beauty of the awareness which is the central mystery of life for each of us.

Physics works in four orthogonal or independent dimensions, x, y, z, and *time*. Because time flows only in one direction, so to speak, it has special properties, expressed in part by relativity. For two or three decades I have tried to indicate the nature of these other aspects of human awareness by saying that they are components extending in a fifth dimension, the esthetic or spiritual dimension, which is orthogonal to x, y, z, and t, and equally real. Here is the locus of such nonmetrical realities as love, beauty, justice, freedom, and honor, and this dimension is as immediate to our perception and as definite a ground for our being as the x, y plane of this floor beneath our feet. This statement is of course only a mode of description, but most scientific men seem to forget that all information is first subjective and immediate, and its objectivity develops only after it has been processed by various operations.

There is nothing original in my remarks to you this evening; all these things have been better said by others. I simply have felt that, as a scientific member of the Cosmos Club I should speak out to those others among us whose activities and competences lie more in the direction of the humanities and the social sciences and the fine arts.

Actually, as you can sense from what I have said, my life experience and my warm conviction is that physics and the natural sciences, especially as they have flowered and enlarged their concerns with human interests during the past half century, are a part of, and participate with vigor in, the humanities and the fine arts, and through philosophy and by their technological off-spring, forcefully condition the social sciences as well. There are materials enough in this one remark for several more quiet discussions among us, but tonight I have wanted only to focus your minds on the liberating quality, for all philosophers and humanists, of the one single contribution made by modern physics in formulating and verifying *complementarity* as a basic property, an intrinsic, inescapable attribute, of the real physical world, insofar as the finite mind of man is ever to be allowed to view reality.

The best expression that I know of these and similar thoughts was given by Robert Oppenheimer in 1953 in a series of Christmas lectures over the BBC in London called the Reith

Lectures, published by Simon and Schuster under the title *Science and the Common Understanding*. I, personally, have long felt deeply grateful to "Oppie" for these and many other insights, ranging all the way from physics to philosophy, and I will share with you a few paragraphs from his beautiful presentation for the BBC. He speaks of the physico-chemical description of living forms and the question whether, despite the electron microscope and radioactive tracers, this kind of description can in the nature of things ever be complete. I quote:

> Analogous questions appear much sharper, and their answer more uncertain, when we think of the phenomena of consciousness; and, despite all the progress that has been made in the physiology of the sense organs and of the brain, despite our increasing knowledge of these intricate marvels, both as to their structure and their functioning, it seems rather unlikely that we shall be able to describe in physico-chemical terms the physiological phenomena which accompany a conscious thought, or sentiment, or will. Today the outcome is uncertain. Whatever the outcome, we know that, should an understanding of the physical correlate of elements of consciousness indeed be available, it will not itself *be* the appropriate description for the thinking man himself, for the clarification of his thoughts, the resolution of his will, or the delight of his eye and mind at works of beauty. Indeed, an understanding of the complementary nature of conscious life and its physical interpretation appears to me a lasting element in human understanding and a proper formulation of the historic views called psychophysical parallelism.
>
> For within conscious life, and in its relations with the description of the physical world, there are again many examples. There is the relation between the cognitive and the affective sides of our lives, between knowledge or analysis and emotion or feeling. There is the relation between the esthetic and the heroic, between feeling and that precursor and definer of action, the ethical commitment; there is the classical relation between the analysis of one's self, the determination of one's

motives and purposes, and that freedom of choice, that freedom of decision and action, which are complementary to it. . . .

To be touched with awe, or humor, to be moved by beauty,
to make a commitment or a determination, to understand some
truth — these are complementary modes of the human spirit.
All of them are part of man's spiritual life. None can replace
the others, and where one is called for the others are in
abeyance.

The wealth and variety of physics itself, the greater wealth
and variety of the natural sciences taken as a whole, the more
familiar, yet still strange and far wider wealth of the life of the
human spirit, enriched by complementarity, not at once com-
patible ways, irreducible one to the other, have a greater
harmony. They are the elements of man's sorrow and his
splendor, his frailty and his power, his death, his passing, and
his undying deeds.

Now I invite all of you as fellow humanists to make
the most of this new and fresh awareness, coming unexpectedly, perhaps, but in most explicit fashion, from the field
we know as physics, this new verification, confirming our in-
tuition of complementarity, inviting and urging us to use whatever modes of questioning and of thought and expression are
best fitted to evoke and to express the spirit of man, as he
perceives the awesome beauty of the world of which he is a
finite part, yet, also, inexpressibly, of infinite extension.

*D*r. *Edwin Hubble wrote "The Nature of Science" as one of the Hitchcock Lectures he delivered at the University of California in 1948. After his death the Huntington Library in California made it the title piece of a collection of Dr. Hubble's non-technical essays.*

Dr. Hubble's contributions to astronomy include some of the great breakthroughs of twentieth-century science. When he began his career in astronomy in 1919, our galaxy — the Milky Way — was thought to be the universe. Hubble focused the new 100-inch telescope at Mount Wilson on the hazy-appearing spiral nebulae and discovered that they consisted of individual stars. By comparing these stars with others whose luminosity and distance from earth were known, he showed that the nebulae are stellar systems like our own. Further study indicated that the universe is composed of billions of such systems.

Hubble also formulated the law of redshifts, which is basic to all present-day theories about the size, shape, and origin of the universe. He found that the spectra of certain stars were out of the normal position — shifted toward the red end of the spectrum — an indication of motion in a direction away from the observer. Later investigation of this phenomenon by Hubble and others resulted in the conclusion that outlying stellar systems are receding from us at speeds increasing with their distance; in other words, that the universe is expanding.

Hubble was born in 1889 in Marshfield, Missouri. His interest in astronomy developed while he was an undergraduate at the University of Chicago, but he took advantage of a Rhodes Scholarship to study law for two years at Oxford University, and was admitted to the bar in Kentucky. In 1914 he left his law practice to study astronomy at the Yerkes Observatory of the University of Chicago. In 1917 he enlisted in the U. S. Army. He was quickly promoted to major, and commanded a battalion in France.

When he returned in 1919 he was recruited for the Carnegie Institution by Dr. Hale, then Director of the Mount Wilson Observatory in California. For the next 22 years Hubble carried on his investigations at Mount Wilson, and later, at Palomar. He died in 1953 at his home in San Marino, California.

Edwin Hubble

THE NATURE OF SCIENCE

From *The Nature of Science and Other Essays,* published by
The Huntington Library, 1954.

The history of modern science began with the Renaissance. It developed slowly at first, then faster and ever faster,
until it now threatens to dominate our civilization. The general
acceptance of the validity of science is a very recent phenomenon. As late as 1880, Huxley complained that "no reply to a
troublesome argument tells so well as calling its author a 'mere
scientific specialist.' " Fifty years later, Belloc complained that
"a thing having been said to be established scientifically, there
is no questioning of it." Today the attitude of the public is even
more "advanced" (to use a dubious adjective). There are many
who assume that the powers of science are unlimited, and
some who refuse to believe a scientist who says that such and
such a thing in his field is impossible.

All of these attitudes reflect a widespread misunder-
standing of the nature of science and, consequently, of its
limitations as well as its capacities. Scientists are well aware of
this confusion as, indeed, are all men with a good liberal education. But these men seldom discuss the matter in public, and
when they do, they rarely reach any considerable audience.

In the past, this reticence was understandable and perhaps even commendable. Scientists in general are not very
articulate; they work in comparative seclusion and they do not
cultivate the art of persuasion. But now a new era has emerged,
and reticence is no longer a virtue. Knowledge has been ac-

57

cumulating at an ever increasing rate, and knowledge, once it is available, can be used for evil as well as for good. It was inevitable that a day would come when the expanding body of knowledge would sweep across the danger level. That day, as you know, has come — and passed. Knowledge is already available by means of which men could wreck the civilization of the world — and the growth of knowledge continues faster than ever before.

It is imperative, in our day, that the applications of knowledge be controlled, and controlled not in one nation alone but throughout the world. This problem is the central problem of our time. A wise solution, enforced, would be the greatest landmark in the history of the human race. If no solution is found, there may be no more history.

The discussion of this paramount problem and its solution should be carried on with a clear understanding of the nature of science and of its applications. Statesmen and citizens alike should use the same language and understand what it is they are saying. Otherwise, the beneficial applications of knowledge may be suppressed along with the dangerous applications, that is, the baby may be thrown out with the bath; or, what is even more important, science may be confused with the applications, and to use another well-worn metaphor, the goose may be killed which lays the golden eggs.

It is for this reason that the scientists are beginning to talk more freely than hitherto. They speak rather haltingly, to be sure, and somewhat diffidently, but they are driven by a sense of urgency. The subject must be clarified, and those who actually practice the discipline should be able to speak with some authority. What they have to say is not entirely new. A few men, including an occasional scientist, have explained the subject fairly well. But others have written nonsense, and the people, if they can, are left painfully to distinguish the good from the bad.

So the scientists are beginning to talk. They will not all say the same things, they will not necessarily agree on details, but if a sufficient number raise their voices, the essential features of the discipline will emerge rather clearly. This clarification of the problem which overshadows our civilization is the first duty of the scientists. The solution of the problem, and its enforcement, is the responsibility of all men — including the scientists, but only in their position as human beings, not as specialists.

For myself, I speak as a student of Astronomy, the type specimen of pure science. It is the oldest of the special disciplines, and it is called the mother of them all. The first notions of law and order in the universe were found in the heavens. When the same ideas were dragged down from the skies to the earth, Physics was born, and then, one by one, the other disciplines followed. But all of the brood were reared in the midst of human society, and all were exposed to the contamination of human desires. The old mother still remained remote and serene, the purest of them all. It is true that certain practical applications of astronomy were devised — timekeeping, navigation, and geography — but these specialties have long since been reduced to routine techniques, and removed from the body of the discipline.

It was not until recently, with the development of spectrum analysis, that this isolation was invaded. Today we look on stars and nebulae as so many celestial laboratories in which atoms can be studied under conditions of temperature and density that far transcend the utmost limits of terrestrial equipment. In the same spirit, the exploration of space may be described as the study of matter and radiation on the grand scale, as distinguished from small-scale studies on the earth. Physics and Astronomy are merging. The first born has come of age, and, indeed, has become head of the family.

There is a unity in science, connecting all its various fields. Men attempt to understand the universe, and they will follow clues which excite their curiosity wherever the clues may lead.

But admitting these interrelations, astronomy may still

be regarded as the purest of the special disciplines, the farthest removed from practical problems of daily life, and within this field we can most readily identify the essential features of science in general.

These features are not mentioned in the dictionaries. There you will find a wide range of definitions of science, and they emphasize the confusion in the language of general discourse. No adequate definition of the word has ever been formulated. The remark of the astronomer is perhaps as good as any of the attempts: "equipped with his five senses, man explores the universe around him and calls the adventure science." It is science in this sense that I propose to discuss — the acquisition of objective knowledge concerning the structure and behavior of the physical universe.

Let me begin with an attempt to emphasize the distinction between science and values, a distinction which is evident in a comparison, for instance, of the laws of motion and the canons of art. The realm of science is the public domain of positive knowledge. The world of values is the private domain of personal convictions. These two realms, together, form the universe in which we spend our lives; they do not overlap.

Positive, objective knowledge is public property. It can be transmitted directly from one person to another, it can be pooled, and it can be passed on from one generation to the next. Consequently, knowledge accumulates through the ages, each generation adding its contribution.

Values are quite different. By values, I mean the standards by which we judge the significance of life. The meaning of good and evil, of joy and sorrow, of beauty, justice, success — all these are purely private convictions, and they constitute our store of wisdom. They are peculiar to the individual, and no methods exist by which universal agreement can be obtained. Therefore, wisdom cannot be readily transmitted from person to person, and there is no great accumulation through the ages. Each man starts from scratch and acquires his own wisdom from his own experience. About all that can be done in the way of communication is to expose others to vicarious experience in the hope of a favorable response.

The distinction between knowledge and wisdom is fully recognized in our time but it was not always so. Men wanted to explore the universe in all its aspects. The attempts began long ago. There was much fumbling; there were many false leads and occasional breath-taking achievements. Eventually it was realized that only one aspect — the world of positive knowledge — could be explored with confidence and, moreover, that success in the venture was measured in terms of disinterested curiosity. Special methods for handling the particular kind of subject matter were developed under the leadership of Galileo and Newton, and modern science was launched upon its extraordinary career.

The requirement of disinterested curiosity was never formulated consciously. Yet it seems to have been the dominant motive in the work of all the great men of science. It has inspired the statement that the essential characteristic of science is the simple idea of attempting "to ascertain objective truth without regard to personal desires."

Men of science, like all other men, spend most of their lives in the larger world of values. There they play their roles as citizens and as human beings. But occasionally they slip out of the circle into another world that knows nothing of values. There they attempt to explore the universe as it is — not as it should be, but as it is. They may not always achieve complete detachment, yet that is their conscious aim. Driven by sheer curiosity, they seek to understand the world — not to control it, not to reform it, merely to understand it. This approach has been extraordinarily successful within a limited field — the field of science. There seems to be no competing attitude in the exploration of new fields of positive knowledge.

The subject matter of science has been described as "judgments on which it is possible to obtain universal agreement." These judgments do not concern individual events, which can be witnessed only by a few persons at most. They are the invariable association of events or properties which are known as the laws of science. Agreement is obtained by observation and experiment — a court of appeal to which men of all races and creeds must submit if they wish to survive.

If anyone refuses to agree with a judgment, we ask him to go and test it for himself. If he still refuses to concur, we ignore his words and watch his actions. There is a story of Simon Newcomb which illustrates the point. A crank rushed into his office one day (Newcomb was Superintendent of the *Nautical Almanac* at the time) and announced belligerently that he did not believe in the law of gravity. Newcomb did not argue the matter; he merely invited the fellow to jump out of the window, and then watched to see what would happen.

The laws of science are derived from, as well as tested by, observation and experiment, and especially from measurement. The measures can never be exact in the absolute sense, and this margin of error must be taken into account in all interpretations of the data. The difficulty is met and overcome by the use of "probable errors," a conception that is peculiar to science. At every stage of an investigation, the uncertainty of each measure and each combination of measures is carefully estimated, and expressed numerically. For instance, a particular measurement is repeated many times and the average of all the results is adopted as the most probable value. Then, from an analysis of the differences in the individual measures, it is possible to say that the error in the adopted mean value has an even chance of being less than, or greater than, a certain quantity. This quantity is called the "probable error."

The conception, as I said, is peculiar to science. It has no place in the world of values. Bertrand Russell has discussed the distinction in his best epigrammatic style. After remarking that subjective certainty is inversely proportional to objective certainty, that we are most certain of those judgments we cannot demonstrate, he reminds us that when a scientist has determined a quantity with unusual accuracy, he is the first to admit that he is likely to be wrong — and he knows about how wrong he is likely to be. And Russell then asks the question, "Who ever heard a theologian preface his creed, or a politician conclude his speech with an estimate of the probable error of his opinion?" These remarks emphasize the fact that science deals with probable knowledge and that the methods of science are adjusted to their proper subject matter; also, they call attention to the fact that a scientist never attempts to speak with personal authority in his field, but merely demonstrates conclusions which anyone (in principle) may verify if he cares to take the trouble.

The laws which are the subject matter of science take many forms, ranging from simple definitions — such as the invariable association of properties which identify, for example, the element of iron — to very complicated associations of events such as the law of falling bodies. However, they are all quite general statements, and they apply to numerous actual or possible cases. The term "invariable" represents an assumption. If an association is observed to hold in many cases of a particular kind, it is assumed that the same association will be found in the next case that will be observed in the future. In carefully controlled experiments the number of test cases may be reduced to a minimum, and occasionally to a single critical trial. The results are confidently stated as laws, merely because they fit into the general pattern of knowledge within the particular field which is based upon innumerable data. Nevertheless, they represent probabilities only. The sun has risen each day in the past. It will probably rise tomorrow, as everyone agrees.

The nature of the subject matter defines the realm of science. The necessity for general agreement restricts the explorations to the field of positive knowledge, and this knowledge concerns not ultimate reality, but phenomena only.

The methods of science may be described as the discovery of laws, the explanation of laws by theories, and the testing of theories by new observations. A good analogy is that of the jigsaw puzzle, for which the laws are the individual pieces, the theories local patterns suggested by a few pieces, and the tests the completion of these patterns with pieces previously unconsidered. Following this analogy, just as local theories are built up from pieces, so general theories are built up from local theories. The scientist likes to fancy, although he cannot demonstrate, that sufficient pieces may be assembled to indicate eventually the entire pattern of the puzzle, and thus to reveal the structure and behavior of the physical universe as it appears to man. At any rate, he has found methods which, when restricted to their proper fields, have been constantly advancing in the desired direction. He is bound to continue the explorations unless or until he runs into a blank wall of self-contradictions.

The laws of science are almost innumerable, but they fall into relatively few general types. The discovery of a new type is a notable event in the progress of science, but once the type is established, lesser men can add individual examples at a rate measured largely by patience and industry.

The laws derived from observation and experiment are discoveries in the true sense of that word, and a new type generally opens to investigation a previously unexplored field of knowledge. Nevertheless, the laws are the subject matter of science; they are not the finished composition. It is theory that integrates the laws, and the making of successful theories is a creative activity whose operation is difficult to understand.

Science is pragmatic. Theories are judged by a single criterion — do they work? Their origin does not matter. They may be inferred, or invented, or dreamed. Until they are tested they are merely working hypotheses; in other words, they are plausible interpretations of data already available. Ingenious men can and do invent many theories to account for a given set of data.

Historians play this game endlessly. Froude remarked eighty years ago that "facts of history are like the letters of the alphabet — you may make them spell what you like." But scientists do not enjoy the same freedom. Each theory, while accounting for the laws already known, predicts new, hitherto unobserved laws. Therefore, the theories of scientists, unlike those of the historian, can be tested by observation and experiment in new fields. And the tests are made. The validity of theories is measured, not by their origins but by the verification of predictions. This procedure is the very essence of the scientific method, and serves to control the powerful but dangerous instrument of inductive reasoning. When theories cannot be tested, their appeal is largely aesthetic.

In these circumstances, the making of theories is a measure of men. A few seem to have a flair for inventing the right kind of theories — the kind that survive the tests, at least for a time. But the majority of men, however ingenious, are handicapped by the inevitable tests.

This feature of theory-making is especially prevalent among theories on the higher levels — among the great generalizations which correlate previously isolated theories and furnish the pattern for very large tracts of knowledge. Universal gravitation waited for a Newton; relativity, for an Einstein.

But even the best of theories are accepted as temporary working hypotheses. The laws of science are relatively permanent, but the theories, which attempt to explain the laws, come and go. They serve their days of usefulness and then they fade away. We know only a little part of the universe, and our knowledge is constantly expanding. We predict what to expect in the new regions, from the information we already possess. When the actual data are found and reported, the theories are always reviewed in the light of the new information. Inconsistencies are pounced upon eagerly because they often point the way to new and broader conceptions. Current theories are discarded, or they are revised or merged into a wider generalization, and the event is welcomed as another step toward the ultimate goal. Such events occur frequently on the low levels of theory when, for instance, the first explorations are rapidly pushed out into newly opened fields. But on the top levels they are rare, as, for instance, when Newtonian gravitation was absorbed into the deeper, more inclusive theory of relativity.

The openings of new fields are exciting periods, and then the methods of science are rather clearly seen. It is like a campaign. Accumulating knowledge piles up at some barrier beyond which lies unknown darkness, the land that challenges the explorer, where, he likes to imagine, almost anything is more than likely to happen. A day comes when a weak spot is found in the barrier, often by chance. Immediately, a breakthrough is engineered, hammered out by every means available. Once the break is made, explorations sweep forward like a flood spreading out until it is stopped by more distant barriers. This first sweep is a reconnaissance of the field, guided by the general knowledge already at hand. Next comes the consolidation of the new country — the careful surveys, the accumulation of new laws, the review of old theories, and, perhaps, the formulation of new theories. Finally, when order is established, the territory is integrated into the main body of knowledge, and theory is reviewed on the high levels.

In principle, the development of a new branch of science begins with observations and simple experiments. The results are reported in terms of measured quantities together with the estimated uncertainties of the measures. These data furnish a growing body of laws which represent the actual subject material. When a sufficient number of laws is assembled, the creation of theories begins. Various laws are grouped together as special cases of a general principle or theory. The work starts tentatively, guided perhaps by analogies in other fields and by a working hypothesis called the uniformity of nature. Most of the theories fail to meet the tests of predictions, but with the accumulating experience of trial and error, men begin to get the feel of the data. A confident, sure touch is developed, and the game becomes exciting. Theorizing on the higher levels follows much the same pattern, until finally the new branch is merged into the general body of science.

These procedures are based on inductive reasoning. From a few scattered cases, we infer a general principle that should apply to all similar cases. The validity of induction has never been demonstrated. It remains as an unsolved problem of logic. Consequently, the theories of science, as well as the measurements, must be accepted on the basis of probabilities. When a theory collapses, an attempt is made to replace it with a more probable theory, and, in this way, science proceeds as a series of successive approximations.

When a sufficient body of high-level theories has been assembled, an entirely new chapter is opened. An attempt is made to derive from the theories a limited set of general principles whose implications cover the entire branch, just as Euclid's axioms were once supposed to contain implicitly the whole of plane geometry. If such an attempt were ever successful, and the chosen set of principles did close and bound the field, that branch of science would be perfected and written off the programs of the explorers. For then the reasoning would be purely deductive, and the ramifications of the field could be confidently predicted without recourse to further observations. Science could then be taught as elementary geometry is taught.

This pleasant speculation suggests the intimate relation between science and mathematics. They are not at all the same thing. Mathematics is not science, although it furnishes the scientist with some extremely powerful tools for the analysis of data and the handling of theories, especially on the higher levels. But mathematics is akin to pure logic. It concerns relations between postulates. Russell, in another happy epigram, has said that the mathematician never knows whether what he is talking about is true, and wouldn't be interested if he did. Haldane once remarked that mathematics deals with "possible worlds," that is, with logically consistent systems, and is not interested in any attribute except logical consistency. Following this lead, it might be said that mathematics furnishes us with a vast array of "possible universes," and that science attempts to identify, among them, the actual universe that we inhabit. The acquisition of new knowledge is constantly reducing the list of possible universes which must contain our own. It is the task of theoretical, or mathematical, physics to keep us informed as to the minimum number, and to indicate critical tests by which the list may be still further reduced.

A conspicuous example of this procedure is found in cosmology, a study which concerns the gross, large-scale features of the physical universe. The combined efforts of mathematical physicists and observers have reduced the array of possible universes to such a limited range that it is now possible to predict with confidence that the type of the actual universe will be identified within the foreseeable future, perhaps within the next decade.

The mathematical physicists are constantly studying the general principles of nature, derived by induction, as though they were postulates which close and bound the fields of science. In the end, the set of postulates must be complete and must be logically consistent. The occasional recognition of inconsistencies has led to reinterpretations of large areas of knowledge, and even to such a major revolution in scientific thought as the theory of relativity.

A set of postulates derived from observations will probably be incomplete; it is unlikely that we know as yet all of the fundamental principles of nature. Theoreticians attempt to measure these gaps in the sets of postulates by the degree to which the field closed and bounded by the postulates fails to coincide with the observable universe. It is possible that new fundamental principles may be found in this way, but their validity must always be checked by observation.

This discussion of the nature of science began with the simple observer, exploring the world about him, and has now wandered into the fringe of theory and mathematical abstractions. If it followed a natural course, it would linger for a while in that field, and then conclude with a dissertation on the philosophy of science. I do not propose to follow this course, but I shall tell you a fable concerning the philosophical aspects of scientific research, before passing on to a summary of the ground that has been covered.

The fable is Eddington's tale of the fisherman — Eddington calls him an ichthyologist — and it will take its place as a classic in the literature of science. An ichthyologist set out to study fish in the ocean. He spent his time dipping a net with two-inch meshes into the water and studying the fish he caught. He made two discoveries, namely, all fish have gills and all fish are more than two inches long. These results represent empirical and *a priori* knowledge respectively. He could not predict the gills but he should have predicted the minimum length. In research, the fisherman's net is represented by the scientist's equipment — his tools, his sensory organs, and his brain which operates in certain patterns. A complete knowledge of this equipment, says Eddington, permits us to predict a large portion of the current body of knowledge — the *a priori* portion — without making a single experiment or observation. The remainder of the body of knowledge — the empirical portion — must be found by the explorers.

Eddington then proceeds to discuss the two kinds of knowledge, and satisfies himself that all of the important and significant knowledge is *a priori,* that even the fundamental constants of nature, from the gravitational constant to the rate of expansion of the universe, can be derived within the study. The empirical knowledge, however curious and interesting, concerns mere trivial details in the scheme of things.

The problem illustrated by the tale of the fisherman is

the central problem in the philosophy of science and the current discussions are highly controversial. Eddington's interpretation represents a minority point of view, but it is presented with such brilliance and persuasive power that it commands respect, if not concurrence.

The explorers are not deeply impressed by the controversy. They are pragmatists, and interpretations are useful only as long as they work, as long as they predict new phenomena and the predictions are verified. They know that in the past most of the really new fields have been opened by the explorers, using the methods of Galileo and Newton rather than the method of Plato. And they will continue their explorations on the assumption that, in the future as well as the past, new fields will be opened which cannot be predicted from the armchair.

From the foregoing descriptions and comments, it is seen that science is necessarily restricted to one aspect of the universe — the objective world of phenomena. It deals with probable knowledge only, its methods are empirical, its philosophy is pragmatic. The scientist explores the world of phenomena by successive approximations. He works in an atmosphere of probabilities; he knows that his data are never precise, and that his theories must always be tested. It is quite natural that he tends to develop healthy skepticism, suspended judgment, and disciplined imagination.

The world of pure values, that world which science cannot enter, has no concern whatsoever with probable knowledge. There finality — eternal, ultimate truth — is earnestly sought. And sometimes, through the strangely compelling experience of mystical insight, a man knows beyond the shadow of a doubt, that he has been in touch with a reality that lies behind mere phenomena. He himself is completely convinced, but he cannot communicate the certainty. It is a private revelation. He may be right, but unless we share his ecstasy we cannot know.

II
THE NATURE OF DISCOVERY

*Merle A. Tuve**

RADIO ECHOES
(THE ORIGIN OF RADAR)

Excerpt from an address at Girard College, Philadelphia,
on receiving the John Scott Award, December 15, 1948.

Before I close this talk, I think I should tell you a little
story, which illustrates how innocent of great things a scientific
man feels at the time of a discovery. I was privileged 23 years
ago to participate in the first discovery of the echoes of radio
waves from the ionosphere, which is the electrically conducting
region 100 miles or so overhead which reflects radio waves
around the world and makes communication possible, instead
of letting the waves fly out into space. Our radio echoes developed into something unexpected a good many years later.

Dr. Gregory Breit, now a Professor at Yale, was a new
member of our Department of Terrestrial Magnetism in the
autumn of 1924, when I was a young instructor at Johns
Hopkins in Baltimore. In 1885 an electrically conducting
layer in the upper atmosphere had been postulated to explain
the daily variations of the compass. Marconi sent his famous
letter "S" across the Atlantic in 1901 and the same conducting layer was the only explanation for receiving his radio waves
in Newfoundland around the curvature of the earth. Dr. Breit
wanted to observe the reflections of short radio waves from
this layer, to be sent upward from a big parabolic reflector in

---

* Dr. Tuve is represented by two articles in this volume. See also
his "Physics and the Humanities" beginning on page 41 and the biographical
note preceding it.

73

Washington, and he asked me to devise the receiving arrangements in Baltimore, 40 miles away.

A meeting of various experts was held at the Department, the problem was considered, and money was approved for a reflector 100 feet high. I waited until the money was in hand, and then confessed to them that I was worried about the difficulties of receiving such short waves (under one meter) at such a distance as Baltimore, after an uncertain degree of reflection by the upper atmosphere. I had previously found it difficult to receive them at a distance of 30 feet.

On the spur of the moment I suggested to the meeting an old and familiar idea, namely, that it might be better to try for echoes instead by sending out short radio pulses, because in this case longer radio waves could be used. Incidentally, my worries were correct; we know now that waves short enough for use with the reflector would have gone right on through the upper atmosphere. At dinner that evening Dr. Breit and I planned the apparatus and saw that the echo idea might explain some other things we knew about, such as poor modulation and rough music at 150 miles from a station which sounded all right when you listened nearby.

Two months after this meeting Professor Appleton in England published a proposal for measuring the height of the ionized layer by slowly varying the frequency of a continuous wave, which experiment he carried out in 1925. Meanwhile, during the winter and in the spring of 1925, Dr. Breit tried some "echo" experiments using broad pulses ("dots") on distant transmitters, observing the received signals on a cathode-ray oscillograph and recording with a Duddell oscillograph. The received signals showed extra peaks, which might have been echoes, but there was no way to be sure that these were not from rough pulses at the transmitter, due to sparks at the keying device.

I came to Washington to spend the summer of 1925 at the Department working with Dr. Breit on this echo problem.

We had decided that the only way to be certain about the echoes was to use vertical reflections, with the transmitter adjacent to the receiver so that we could be sure by oscillographic observation that the transmitted pulses were clean while we simultaneously observed the extra pulses, due to echoes, on the receiver.

We persuaded Dr. A. H. Taylor at the Naval Research Laboratory to let us apply pulse modulation to a Navy transmitter, and we installed our receiver and oscillographs there.

On July 10, 1925, we first observed echoes, using 80-meter waves. They were very clear, and the transmitted pulses were without extra pulses. However, the distance indicated by the echoes corresponded to the distance to the Blue Ridge Mountains, and it remained constant. This continued for many days. We were not after echoes from distant objects, we wanted to study the upper atmosphere, and we felt disappointed. Ten days later, however, we were able to use the transmitter during the evening hours, and we saw that the distance had changed; in fact it changed still more while we observed. We then knew that at least some of our echoes were from the upper atmosphere.

With our several colleagues we continued to observe echoes and improve the experiments during 1926 to 1928. We made pulses with a duration of 2/10,000 second and separated the multiple echoes. We worked at various shorter wavelengths.

During one of our experiments, which we called the "echo-interference experiment," we were greatly troubled by planes taking off and landing at an airport two miles away. We were not interested in airplanes, and had to wait for the air to clear. We also saw various other transient echoes and effects which we were not interested in. We talked freely about our experiments, and published the echo measurements. Experiments with very short waves, with movable parabolic reflectors, were considered, too, but this was not feasible because no vacuum tubes were then available for transmitters at such short wavelengths. In 1929 we persuaded the Bureau of Standards to adopt the echo method, and we stopped our experiments for three years.

The idea of using radio waves reflected from planes and ships as a military device was privately recognized and made a Navy secret by Lieutenant W. S. Parsons at the Naval Research Laboratory in 1932, seven years after our first experiments there. Today he is Admiral Parsons, the atomic bomb expert.

Professor Appleton and others in England adopted our pulse technique for upper-atmosphere studies in 1929, and carried it forward with results which surpassed our own studies. In 1934, five years later, the British military services independently saw the values of pulse-radio echoes from planes and ships, and quietly made the idea their secret. When the United States and Britain exchanged secrets in 1939 or 1940, both sides were then surprised. At about that time, roughly 15 years after our first echo observations and our disappointment and concern about echoes from the Blue Ridge Mountains, a British physicist devised a vacuum tube which would generate pulses of high power from very short wave transmitters, and the use of movable parabolic reflectors became feasible.

Radar, a simple application of pulse-radio technique and the observation of timed echoes, was a slow but direct outgrowth of our experiments on the upper atmosphere, and was probably the most important single technical device used in World War II. The patent lawyers of RCA cited our early published experiments in preventing the patent offices of various governments from granting anybody a basic patent on radar. Dr. Breit and I were naturally pleased by all these developments, and glad that our part in it was a gift to the public from research activities in pure science.

When Dr. George Harrison Shull presented his paper "A Pure-Line Method in Corn Breeding" to the American Breeders Association in 1909, he was disclosing a technique that in the next generation was to revolutionize the raising of corn, add billions of dollars to the American economy, and significantly increase food supplies throughout the world.

Shull used corn (maize) to demonstrate that self-fertilization results in a weaker strain, whereas successive crossbreeding of two pure strains results in an improved strain with "hybrid vigor." He also showed how crossbreeding would enable a farmer to produce corn plants of almost any desired size or type. The application of Shull's discoveries to corn-growing now adds an estimated 3 billion dollars annually to the value of this crop in the United States alone.

George Harrison Shull was born in Ohio in 1874. He was graduated from Antioch College in 1901, and received his Ph.D. in botany and zoology in 1904 from the University of Chicago. While still a graduate student he served the U. S. Bureau of Plant Industry as a botanical expert.

At about this time Shull became interested in what is now known as genetics, but the term had not yet been invented in 1904. From 1904 to 1915 he was associated with the Carnegie Institution at its Station for Experimental Evolution at Cold Spring Harbor, Long Island, New York, which is now known as the Genetics Research Unit. He began his research by studying a large number of different plants and animals, in search of suitable genetic research material. He finally settled on corn as the most useful for his purposes, and this proved to be a most fortunate choice. At the Genetics Research Unit, corn is still used as one of the materials of genetics research.

While Shull was with the Institution, he also undertook a study of Luther Burbank's methods of plant breeding, which were at that time very famous and had stirred considerable controversy. In 1915 he joined the faculty of Princeton University, where he was professor of botany and genetics until 1942, and professor emeritus until his death in 1954.

78

George Harrison Shull

A PURE-LINE METHOD IN CORN BREEDING (HYBRID CORN)

From *American Breeders' Association,* May 1909.
By permission of the *American Genetic Association.*

Last year I described* a series of experiments with Indian corn which led me to the conclusions: (1) that in an ordinary field of corn the individuals are generally very complex hybrids; (2) that the deterioration which takes place as a result of self-fertilization is due to the gradual reduction of the strain to a homozygous condition; and (3) that the object of the corn-breeder should not be to find the best pure-line, but to find and maintain the best hybrid combination.

The continuation of these studies during the past year has given still further proof of the correctness of the first two of these propositions, and besides has given unexpected suggestions for a new method of corn breeding by which the essential feature of the third proposition may be realized. It is my purpose to discuss this new method briefly in the following pages.

I will first, however, describe the results of the past year's experiments in so far as they bear upon the points in which we are interested here. For convenience I will refer to the two self-fertilized families contrasted in my paper last year as "Strain A" and "Strain B." It will be remembered that these

---

* The Composition of a Field of Maize. Report American Breeders' Association, 4: 296-301, 1908.

79

two families resulted from the self-fertilization of different, apparently equal, individuals; but that notwithstanding this fact, they differed from each other in height and stockiness of stems, width and greenness of the leaves, length of shank of the ears, appendages of the husks, quality of the grains, and the number of rows of grains on the ears. (See Fig. 1.)

In addition to the parallel cultures of self-fertilized and cross-fertilized families which have been continued from the beginning of these experiments in 1904, I had during the past season the F_1 offspring of a cross between two sibs in Strain A, and two families representing reciprocal crosses between Strain A and Strain B. It was observed that every one of the mentioned characteristics which distinguished Strains A and B, remained constant distinguishing features in the pure-bred families, but in regard to the number of rows on the ears, it is now obvious that Strain A has the normal mean number 8, as compared with 14 in Strain B, for in this year 89 per cent of the ears produced by Strain A had only 8 rows of grains, though the selection of ears for seed in this strain during three years was for 12 rows on the ear, and only in the last year was an 8-rowed ear used because a suitable 12-rowed ear was not available. This result is a striking confirmation of the suggestion made last year that according to the law of regression the occurrence of a mean number of rows less than 12 in Strain A indicated that the normal number of rows for this strain is 10 or possibly only 8.

The cross between two sibs in Strain A was grown beside the self-fertilized family belonging to the same strain, and these two families were so similar during the entire period of their development that they were considered identical, but at the end of the season it was found that the cross-bred family was a trifle taller and produced over 30 per cent more grain by weight than the self-fertilized family. In the self-fertilized family, 73 ears were produced, weighing 12 pounds, and in the cross between sibs the 78 ears weighed 16½ pounds. There

Fig. 1. Typical specimens of Strain A (at right) and Strain B (left), showing contrast of vegetative characters. Drawn by J. Marion Shull from a photograph.

was also a striking difference between these two families as regards variability in the number of the rows on the ear, as may be seen in this table:

NUMBER OF ROWS ON THE EARS	8	10	12	14
Self-fertilized	65	6	2	0
Cross-fertilized	8	50	19	1

Unfortunately the parents of these two families were not identical in the number of rows, the mother of the self-fertilized family having 8 rows and that of the cross-fertilized family 10. The greater height, greater weight of grain produced, the higher number of rows on the ears, and the greater variability in the number of rows in the cross-fertilized family all point to the same conclusion, namely, that my self-fertilized Strain A was not yet reduced completely to a homozygous condition, and that the parents, or at least one of them, of my cross-bred family was heterozygous.

The two families which were the product of reciprocal crosses between Strain A and Strain B have proved of great interest, for although the individuals of both Strain A and Strain B were small and weak, and the self-fertilized families of these produced respectively only 12 pounds and 13 pounds of ear-corn, the hybrid family in which Strain A supplied the mother and Strain B the father produced 92 ears weighing 48 pounds, and the reciprocal cross produced 100 ears weighing 55 pounds.

If we reduce these results to bushels per acre on the basis of 10,000 ears per acre and 70 pounds per bushel, it is found that Cross A×B has produced 74.4 bushels per acre and Cross B×A has produced 78.6 bushels per acre, the average for the two families being nearly 77 bushels per acre. The two families which I have kept continuously cross-bred during the period in which these experiments have been in progress, and which have been likewise continually selected to 12 and

14 rows of grains, may be properly taken as controls. These two families together produced 203 ears weighing 107½ pounds, or at the rate of 75 bushels per acre, and when the comparison is extended so as to include my other continuously crossed families — 8 families in all — it is found that these produced collectively at the rate of a little less than 75 bushels per acre.

My farmer friends in the heart of the corn country will not be greatly impressed with these yields of 75–78 bushels per acre, but I must call attention to the facts that the light gravelly soil of Long Island bears a very unfavorable comparison with Mississippi valley alluvium for the production of Indian corn, and further that the summer of 1908 was notable for one of the longest periods without rain that has ever been experienced there.

The important point will not be missed however that the crosses between two self-fertilized strains yielded a little more grain than those strains which had been kept carefully cross-fertilized by hand. To be sure, the difference is not great enough to seem of any particular significance in itself, but it must be remembered that the two self-fertilized strains, A and B, have been essentially unselected, being simply those two strains which have first approached the pure homozygous state as a result of self-fertilization. It is scarcely conceivable that other pure strains crossed together should not give in certain combinations considerably greater yields than those produced by the combination of Strains A and B. At any rate the result is sufficiently striking to suggest that the method of separating and recombining definite pure-lines may perhaps give results quite worth striving for.

This suggestion will be more readily appreciated perhaps if I discuss briefly the theoretical aspect of this method of pure-line breeding as compared with the method now in use among the most careful corn breeders. In the light of my results, the constant precautions that are taken in the method now in use to prevent in-breeding, have for their real object the retention of the most efficient degree of heterozygosis or hybridity, and it is obvious that the selection of the most vigorous individuals for seed really picks out those individuals which have this most efficient degree of hybridity. While I have not investigated the inheritance of the various characteristics of the pure lines of maize and am not in a position to say that they all follow Mendel's law, many investigations of particular characteristics in corn have shown that those characteristics are Mendelian. Even if some of the differentiating characteristics of corn should not prove to be Mendelian, it seems not improper to discuss the two methods on the Mendelian basis.

In the method which selects for seed the most heterozygous individuals, the characteristic splitting and recombination of unit-characters must produce an offspring of quite various degrees of heterozygosis. Some individuals will be as complex as the selected parents, others will have many of the same units in the homozygous condition, and thus be less complex and consequently less vigorous. According to the laws of chance a few individuals in the field may be expected to be almost or quite completely homozygous, and as a result will be very inferior in vigor and will produce but little grain. The result of such a process must always be to give a crop of lower average yield than the average of the selected seed. Moreover, these different combinations of unit-characters and different degrees of hybridity in the offspring of a complex hybrid must introduce a certain amount of heterogeneity into the crop which will have the effect to also lower the average quality with respect to any other desirable points which have been used as guides in the selection of the seed-corn, and efforts at the attainment of homogeneity by the method now in use tend to lessen physiological vigor, and therefore lessen the yield, owing to the fact that such homogeneity in the offspring of hybrids is to be attained only through homozygosis in respect to all those characteristics which affect the form and size of the ear, width, depth, shape, and composition of the grains, and any other feature in which homogeneity may be desired. This is doubtless the explanation of the interesting experience related by Mr. Joseph I. Wing at the meeting of the American Breeders' Association in Columbus two years ago. His father had selected a very fine deep-grained variety of corn in which great uniformity had been attained but only at the expense of decreased yield.

In the pure-line method outlined below all individuals in the field will be F_1 hybrids between the same two homozygous strains, and there are theoretical grounds for expecting that both in yield and uniformity superior results should be secured. Thus, every individual will be as complex as every other one and should produce an equal yield of grain if given an equal environmental opportunity, so that in so far as hereditary influences are concerned the vigor of the entire crop should be equal to the best plants produced by the methods now in use. This would seem to result necessarily in a larger yield than can be produced by the present method. But not only will all the plants in the field have the same degree of complexity, but they will all be made up of the same combination of hereditary elements, and consequently there must result such uniformity as is at present unknown in corn.

With such a prospect as this, I believe we will be sufficiently interested to make the discussion of the method by which such results are to be attained worth while. The question naturally arises as to whether the technique of the new method will be sufficiently simple to make it practicable. To this question I believe I can safely answer that the pure-line method will be considered simpler than the elaborate one now in use among the most careful breeders, *e.g.*, those at the Illinois, Connecticut, and Ohio State Experiment Stations. The process may be considered under two heads: (1) finding the best pure-lines; and (2) the practical use of the pure-lines in the production of seed-corn.

(1) In finding the best pure-lines it will be necessary to make as many self-fertilizations as practicable, and to continue these year after year until the homozygous state is nearly or quite attained. Then all possible crosses are to be made among these different pure strains and the F_1 plants coming from each such cross are to be grown in the form of an ear-to-the-row test, each row being the product of a different cross. These cross-bred rows are then studied as to yield and the possession of other desirable qualities. One combination will be best suited for one purpose, another for another purpose. Thus, if the self-fertilized strains be designated by the letters of the alphabet, it may be found that Cross C×H will give 120 bushels per acre of high-protein corn, that F×L produces a similar yield of low-protein corn, that K×C gives the highest oil content accompanied by high yield, and so on. Moreover, it seems not improbable that different combinations may be found to give the best results in different localities and on dfferent types of soils. The exchange of pure-bred strains among the various experiment stations would greatly increase the number of different possible hybrid combinations and facilitate the finding of the best combination for each locality and condition.

(2) After having found the right pair of pure strains for the attainment of any desired result in the way of yield and quality, the method of producing seed-corn for the general crop is a very simple though somewhat costly process. Two isolated plots will be necessary, to which I may conveniently refer as Plot I and Plot II. (See Fig. 2.) In Plot I will be grown year after year only that pure strain which investigation has proved to be the best mother-strain for the attainment of the desired end. Thus, if it has been found, as in the example already cited, that Cross C×H gives the desired result, Plot I will be occupied by Strain C. This will require no attention from the breeder's point of view except that any exceptionally vigorous or aberrant individuals should be eliminated, as such plants might be safely assumed to be the result of foreign pollinations. In Plot

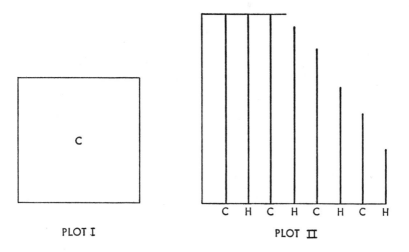

Fig. 2. Arrangement of the two isolated plots for the pure-line method of corn breeding.

II Strain C and Strain H are to be planted in alternate rows, and all of Strain C is to be detasseled at the appropriate time. All the grain gathered from the detasseled rows will be seed corn for the general field crop, and that gathered from the tasseled rows will be pure-bred Strain H to be used again the following year in the same way. Here again in pure Strain H all exceptionally vigorous or aberrant individuals should be discarded as being probably due to the entrance of foreign pollen.

I am not prepared at present to say what will be the probable cost of seed-corn when produced by this method, but have reason to suppose that it would be more expensive than by the present method; nor can I surmise what relation this increased cost will bear to the increased yield that will be produced. These are practical questions which lie wholly outside my own field of experimentation, but I am hoping that the Agricultural Experiment Stations in the corn-belt will undertake some experiments calculated to test the practical value of the pure-line method here outlined.

In "Pulsating Stars and Cosmic Distances," written for Scientific American *of July 1959, Dr. Robert Kraft provides a clear and simple explanation of the way in which astronomers use these strange celestial objects to compute the distances between stars.*

Dr. Kraft's special fields are stellar spectroscopy and galactic structure. His first major investigation while a member of the staff of Mount Wilson and Palomar Observatories was a study of the SS Cygni stars, a group of explosive variables which he showed were all close binary systems. He predicted that in such a close pair of stars, the brighter partner would gradually grow larger, exchange some mass with its fainter brother, and lose some mass into space.

From his observations of binary systems like SS Cygni as well as a number of old novae, Kraft deduced that their irregular variability and explosive changes were caused by such an exchange of mass. His most recent work at Mount Wilson and Palomar has been on stellar rotation.

Robert Paul Kraft was born in Seattle, Washington, in 1927. He was graduated from the University of Washington in 1947, and earned his master's degree there in 1949. A fellowship at the Lick Observatory in 1952 and 1953, and a National Science Foundation fellowship from 1953 to 1955, enabled him to return to his graduate studies, this time to the University of California at Berkeley, where he received his Ph.D. in astronomy in 1955.

Dr. Kraft spent the next year at Mount Wilson and Palomar Observatories as a National Science Foundation postdoctoral fellow. He was assistant professor of astronomy at Indiana University from 1956 to 1958, and at the University of Chicago from 1958 to 1960. Since 1960 he has been on the staff of Mount Wilson and Palomar Observatories.

Robert P. Kraft

PULSATING STARS AND
COSMIC DISTANCES

From *Scientific American*, July 1959.

Our present picture of the universe — its structure, size and age — rests to a large extent upon observations of a few pulsating stars. Each of these stars waxes and wanes as much as one full magnitude (2.5 times) in brightness according to a fixed rhythm ranging in period from less than a day to more than 50 days. In general, the longer the period, the greater the luminosity of the star. Such stars are called cepheid variables after their prototype, star delta in the constellation Cepheus; the most familiar of them is Polaris (the pole star), which brightens and fades in a period of 3.97 days. We do not know what causes the pulsation of cepheid variables, nor what it signifies in the biography of a star. Some 40 years ago, how-ever, by a bold stroke of invention, the variable luminosity of these stars was made to furnish a distance scale that gives astronomy its reach into the cosmos beyond the immediate neighborhood of the solar system.

The new distance scale at once made it possible to locate the center and to measure the dimensions of our galaxy. A few years later the presence of cepheid variables in celestial objects such as the Great Nebula in Andromeda helped estab-lish that these "nebulae" are themselves galaxies — island universes as large as our own located at immense distances out in space. But in recent years the profound usefulness of the cepheid distance scale has been almost overshadowed by its defects. Corrections in the scale have made it necessary for the dimensions of the observable universe outside our galaxy to be doubled, and still further revisions may be required. Because the age of an expanding and evolving universe can be deduced from its distance scale, cosmologists have concurrently had to revise the age of the universe upward, from two billion to perhaps 10 billion years. These corrections and further refinements still in progress derive from closer study of the cepheids themselves. It now seems safe to say that the cosmic distance scale will not again expand so radically, and that it is at last ready for secure calibration.

In all likelihood we shall achieve this objective still without understanding why the cepheids pulsate. Among the 15,000 stars listed in the monumental new Soviet *Variable Star Catalogue,* edited by B. V. Kukarkin, P. P. Parenago, Y. Efremov and P. Kholopov, about 3,000 exhibit the regular pulsation of the cepheids. Spectroscopic observation shows that the surface temperature of these stars varies upward and downward in phase with their light. Apparently they also expand as they brighten and contract as they fade. In the 1920's Sir Arthur Eddington was able to show theoretically that the rate of pulsation must be related to the mean density of the cepheid (its mass divided by its volume), much as the period of a pendulum on earth is governed by its length. But we have no mechanism to explain this behavior, and we cannot say why a star becomes a cepheid.

The most important advance in our knowledge of the cepheids — and the most drastic revision of the distance scale — came a decade ago with Walter Baade's discovery that the stars of the universe may be divided into two major populations. To Population I, made up of young, hot, short-lived stars, he assigned the brighter and longer-period cepheids that appear in the arms of spiral galaxies. The fainter and shorter-period cepheids associated with the globular clusters that swarm around the centers of galaxies Baade placed among the older and longer-lived stars of Population II. While astronomers now believe that Baade's two populations represent an oversimplification and that stars are more continuously graded in age, the cepheids seem mostly to belong to the extreme ends of the population spread.

At present we imagine that the young Population I cepheids represent a phase in the life of any star. If we plot the color (that is, the temperature) of stars against their absolute luminosity (their intrinsic brightness corrected for distance), most of them occupy a rather well-defined "main sequence." To the right of the main sequence is a scattering of other stars, most of them "red giants." Between the main sequence and the red giants is an "instability strip" containing the cepheids. We presently conceive that a star starts out bright and hot, after a very rapid stage of gravitational contraction; then, after the star has consumed a certain amount of its hydrogen fuel, it begins to cool. Thus in terms of the color-luminosity diagram a star spends most of its life on or near the main sequence, but eventually evolves to the right. When it reaches the instability strip, it begins to pulsate. As the star passes through this strip, in the course of a few million years, its pulsation slows and lengthens in period. Upon reaching the end of the strip it ceases to pulsate and becomes a red giant. Ultimately it dims into the graveyard of the white-dwarf stars.

This hypothetical account does not, however, cover the evolution of the old Population II cepheids found in globular clusters. Perhaps these enter the instability strip by evolving "backward" from the red-giant phase instead of from the main sequence. Most of the globular-cluster cepheids have very short periods of less than a day, but even those having longer periods can be clearly distinguished from Population I stars of similar period. Long-period globular-cluster cepheids are on the average 1.5 magnitudes fainter than the younger long-period cepheids, exhibit quite different spectra and have masses only about a fourth as large.

The cepheids are highly luminous stars. Polaris, the nearest of them, is not a particularly bright cepheid, but it is about 600 times brighter than the sun. The brightest Population I cepheids are more than 10,000 times more luminous than the sun! This is a fortunate circumstance so far as the measurement of extragalactic distances is concerned, because it means that such stars make themselves visible at very long range.

In order to understand how pulsating stars can furnish a distance scale, we must go back 50 years to the work of Solon I. Bailey and Henrietta S. Leavitt of the Harvard College Observatory. Bailey carried out an extensive investigation of the cepheids in globular clusters within our own galaxy. He found that almost all had periods of less than a day, except for a few that had periods in the range of 12 to 20 days. Miss Leavitt later studied the cepheid variables that appeared in great numbers in photographs of the Clouds of Magellan, the two small galaxies that are companions of our own; she found that most of these cepheids had periods of more than a day. Even more remarkable was Miss Leavitt's discovery that the average apparent brightness of the Magellanic Cloud cepheids is directly correlated with the length of their respective periods of pulsation. Bailey had found no such dependence of luminosity on period in the globular-cluster cepheids, at least those with a period of less than a day.

Astronomers soon recognized the promise of Miss Leavitt's finding. It was known even then that the Magellanic Clouds are distant congregations of stars. Thus the cepheids in the Clouds are all at virtually the same distance from the solar system, and the light of all is attenuated to the same extent by its journey to the earth. Miss Leavitt's measurements of the varied apparent brightness of these stars could therefore be taken as indications of their relative absolute brightness. Here was a potential yardstick for measuring really long distances in the universe!

It was obvious that, if the distance of the Magellanic Clouds could be ascertained, one could determine the absolute brightness of the cepheids. Miss Leavitt's period-luminosity scale could then be used to find the distance to any stellar system or subsystem containing cepheids by turning the problem around: Measure the period of the cepheid, read off its absolute luminosity from the period-luminosity scale, compare this with the observed apparent luminosity of the cepheid and find the star's distance by applying the law that the intensity of light varies inversely with the square of the distance. Of course the accuracy of such a measuring rod depends on the assumption that cepheids in all parts of the universe obey the same period-luminosity law Miss Leavitt had derived from the cepheids in the Magellanic Clouds. This turned out to be a pivotal assumption.

At the time of Miss Leavitt's discovery there was unhappily no way to ascertain the distance of the Magellanic Clouds. Stellar-distance measurement still depended on direct trigonometric parallax, which is effective only for nearby stars. Against the background of stars distributed in the depth of space at all distances from the sun, a nearby star appears to shift its position as the earth travels from one side of the sun to the other. It is thus possible to measure the distances of such stars by simple trigonometry. Even these distances are so large that it is convenient to describe them with a unit called the parsec. We say that a star is at a distance of one parsec if its parallax, that is, half its shift of position, equals one second of arc. But the nearest star has a parallax of slightly less than .8 second of arc. This corresponds to a distance of slightly more than 1.3 parsecs, or 25,000 billion miles. Sirius, the brightest star in the sky, is 2.7 parsecs away, and the parallax of a star at a distance of 100 parsecs is only .01 second. Such small angles cannot be determined very precisely; a distance of about 30 or 40 parsecs is the practical limit for determination by direct trigonometric means.

The cepheids are so rare in space that the nearest of them — Polaris — is 90 parsecs away. It is clear, therefore, that trigonometry could not be used to determine the distance of a single cepheid, and could yield no information on the absolute brightness of even the nearby cepheids.

How, then, could the distance to any cepheid be obtained? Before Miss Leavitt had made her discovery, astronomers had devised a method for measuring what might be called the "middle distances" of our galaxy. With so many stars on our photographic plates we may assume that many stars in any given group have the same absolute brightness. We may also assume that the motions of these stars, either radially in the line of sight or transversely across the sky, will be at random. Now with the spectrograph we can determine the actual radial velocity of any observable star, independent of its distance from us. The spectrum is shifted toward the violet if the star is approaching and toward the red if it is receding, and the extent of shift gives us the velocity of its motion. On the other hand, the apparent transverse motion across the line of sight (called the proper motion) does depend on distance. If the stars of our given group move, on the average, with the same actual velocities independent of distance, then the proper motions of these stars will appear to get smaller with distance. Of course relatively few stars are near enough to the sun to have exhibited any proper motion during the first century of photographic astronomy. But when we have determined the statistical spread of the radial velocities, it is reasonable to suppose that the proper motions vary in the same range. Since the distribution of proper motions does decrease with distance, the identification of the spread in radial velocities with the spread in proper motions indicates the average distance to the group of stars under consideration.

With the mean distance obtained in this way, one can correct the mean apparent magnitude of the stars for the effect of distance and get the average absolute magnitude. From studies of this sort in 1913 Ejnar Hertzsprung of Denmark found an average absolute magnitude of —2.3 for a cepheid with a period of 6.6 days. (On the magnitude scale the lower number refers to the brighter star; stars brighter than the first magnitude have negative magnitudes.) Hertzsprung's result was based on only 13 nearby cepheids for which the proper motions were known. But astronomers now had the absolute luminosity value needed to convert the apparent luminosity of any cepheid to absolute luminosity by reference to Miss Leavitt's period-luminosity scale.

In 1918 Harlow Shapley of the Harvard College Observatory saw how the scale could be applied to determine the distances of the globular clusters in our galaxy. He fitted the long-period cepheids (periods of 12 to 20 days) of the globular clusters to the period-luminosity scale for the cepheids of the Magellanic Clouds. From this he determined the absolute luminosity and hence the distance of the long-period cluster stars. Using this determination of the distance to the clusters, he deduced that the mean absolute magnitude of the numerous fainter cepheids in the clusters with periods of less than a day was a little brighter than zero (*i.e.,* some 100 times brighter than the sun). Shapley then had a scale to measure the distance to the clusters that contain only faint, short-period cepheids. From the globular-cluster distances thus derived, he deduced that the globular-cluster system was centered on a point about 16,000 parsecs from the sun in the direction of the constellation of Sagittarius. It seemed reasonable to identify this point with the center of our galaxy. Shapley had obtained the first good estimate of the size of any galaxy. Later determination of the luminosities of these shorter-period cluster cepheids, obtained by proper-motion and radial-velocity studies, have verified Shapley's deduction and shown his estimate to be of the right order.

The period-luminosity scale could also be used to estimate the distances to any nearby galaxy that contains cepheids.

Edwin P. Hubble and his associates at the Mount Wilson Observatory soon ruled off the distance to the Magellanic Clouds and to the Great Nebula in Andromeda. By the comparison of apparent to absolute magnitude thus effected for these and other more distant galaxies, the cepheid distance-scale made it possible to calibrate the spectrographic shift toward the red for the measurement of distances to the throngs of even more distant galaxies so faint and tiny that the cepheids and other stars in their populations cannot be resolved. Cosmologists working from these data were able to estimate the size of the universe and its age from the time of its initial expansion. All this extrapolated from the observation of the peculiar process of cepheid pulsation that we do not yet fully understand!

In the next 25 years, however, astronomers and cosmologists encountered numerous difficulties that cast increasing suspicion on the period-luminosity relationship upon which the whole edifice was built. All other galaxies, as measured by the cepheid distance-scale, were smaller in size than our own, a peculiarly self-aggrandizing result. As nuclear physicists succeeded in calibrating the rate at which uranium and thorium have been decaying to lead in the rocks of the earth, their "clocks" made the earth appear considerably more ancient than the universe. There was difficulty also in reconciling Eddington's calculation of the mean density of the cepheids with density estimates derived from the relationship of the observed luminosity of these stars to their rate of pulsation.

An observation by Hubble and Baade finally opened the way to a test of these suspicions. They pointed out that, if the distance to the Andromeda Nebula had been correctly measured, then the brightest stars of the globular clusters surrounding its central region appeared to be too faint compared to the brightest stars in the globular clusters of our own galaxy. If these bright stars in the Andromeda Nebula were assigned the same absolute brightness as the corresponding stars in our galaxy, then the cepheids visible in the Andromeda Nebula and many of the longer-period cepheids in our own system would also have to be assigned a higher absolute magnitude with respect to the shorter-period cepheids of the globular clusters that had formed the basis of Shapley's scale. Could it be that the globular-cluster cepheids obeyed a period-luminosity law different from that observed for other cepheids?

Such a possibility was foreshadowed in 1940 by an observation made by Alfred H. Joy at the Mount Wilson Observatory. He found a marked difference between the spectrum of a 15-day cepheid in the vicinity of the solar system and a 15-day cepheid in a globular cluster. Then, during the war years, Baade was able to devote the 100-inch telescope on Mount Wilson almost full time to his study of the stellar populations in the Andromeda Nebula. In dividing all stars into two populations he also found a basis for classifying the cepheids into two species.

With the 200-inch telescope in operation on Palomar Mountain shortly after the end of the war, Baade set out to observe the two types of stars "side by side," that is, at the same distance. Unfortunately not even the 200-inch telescope can resolve the faint short-period cepheids in the globular clusters of the Andromeda Nebula. But Baade was able to measure the Population I cepheids of that galaxy with great accuracy against the brightest globular-cluster stars, for which absolute magnitude had been established with the help of the Population II cepheids in our galaxy. Shapley had set the absolute magnitude of these stars at —1.5, based upon his determination that the shorter-period cluster cepheids have an absolute magnitude of zero. The distance to the Andromeda Nebula, calculated from its Population I cepheids in accord with the established period-luminosity scale, predicted that the bright globular-cluster stars should have an apparent magnitude of 20.9. Baade found that these stars were actually magnitude 22.4. In other words, they were 1.5 magnitudes fainter.

This demonstrated that the estimate of the distance to

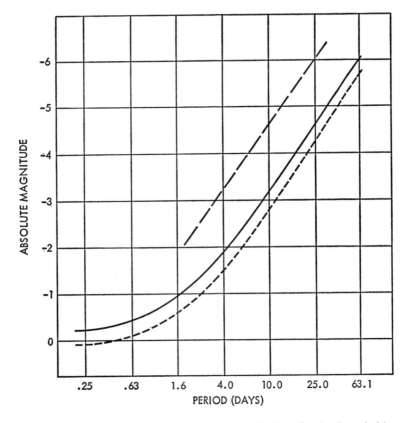

Period-luminosity relation used by Harlow Shapley fitted all cepheids into one curve (middle), with the short-period cluster variables at the lower end. The period-luminosity relation of Walter Baade divides the cepheids into Population I (top) and Population II (bottom). The latter stars are fainter than Population I stars of the same period. On magnitude scale, brightness increases by a factor of 2.5 from −1 to −2, and so on.

Andromeda was too small by a factor of about two. It also showed that the absolute brightness of the Population I cepheids in the Andromeda Nebula was 1.5 magnitudes brighter than had been indicated by the period-luminosity scale. They have a lower apparent magnitude because they are farther away than had been supposed. With distance to the Andromeda Nebula doubled, its size also doubled, bringing it into line with the size of our own galaxy. These results were dramatically confirmed when A. D. Thackeray and A. J. Wesselink of the Radcliffe Observatory in South Africa discovered short-period cepheids in the Large Magellanic Cloud at exactly the magnitude predicted by Baade.

Hindsight now fully explains the discrepancy in the period-luminosity scale. With the Population I cepheids advanced 1.5 magnitudes in luminosity, there is a discontinuity in the scale that clearly divides the cepheids into two types. We also understand why this distinction was missed in the early part of this century. The young Population I stars in the arms of our spiral galaxy lie close to its central plane; the brighter light of these stars is accordingly dimmed by the clouds of dust and gas in which stars are formed. The older Population II stars, which resemble the stars in globular clusters, have had time to drift above and below the galactic plane, so their dimmer light reaches us without obscuration. By a remarkable coincidence the interstellar absorption of the light from the Population I cepheids almost exactly equals the difference in the actual brightness of Population I and Population II cepheids, that is, 1.5 magnitudes. No such obscuration dims the light of Population I cepheids in the Andromeda Nebula or the Magellanic Clouds; their lower apparent magnitude is now correctly attributed to their greater distance. Thanks to this combination of circumstances Shapley was able to fit the long-period cepheids in globular clusters to Miss Leavitt's period-luminosity curve for the cepheids in the Magellanic Clouds. He could not have known that the two types of stars are quite different objects.

From the time of Miss Leavitt's first observations the distinction between the two species of cepheids had also been obscured by a scatter of about one magnitude in the positions of the stars along the mean line of the period-luminosity curve.

For many years this was attributed to observational error and possibly to internal absorption within the Magellanic Clouds. But the scatter could also result from a bona fide physical departure of a given star from the mean line. This is a point of more than academic interest; such uncertainty in the magnitude of a particular star corresponds to a factor of 50 per cent in the computation of its distance. The range of error is too great if the objective is to measure the distance to a galaxy in which only one or two cepheids are available. Accurate determination of distances to individual cepheids has also assumed new importance in the study of our own galaxy. Population I cepheids might be expected to outline the spiral arms of our galaxy and, being very luminous, to carry our knowledge of the spiral structure to considerable distances from the sun.

We are now certain that the scatter is real. Highly accurate photoelectric measurements of cepheids in the Small Magellanic Cloud by Halton C. Arp of the Mount Wilson and Palomar Observatories have established that the scatter is very much larger than the errors of observation. Allan R. Sandage of the same observatories has offered an explanation. Sandage predicts from the theoretical period-density relation that the period-luminosity law must be amended to take account of a third variable. This variable is the surface temperature of the star.

Observations of certain cepheids for which highly accurate surface temperatures and absolute magnitudes can be derived seem to confirm Sandage's theory. These stars are members of loose clusterings of very young Population I stars in our galaxy called open or galactic clusters. The first two were found by John B. Irwin in 1955 at the Radcliffe Observatory. Others were located by Sydney van den Bergh and myself, and the number of such cepheids is now about 10. Their colors (hence surface temperatures) and absolute luminosities are obtained by yet another method for determining distances to stars. We may expect stars that are close together on the color-luminosity diagram, and thus are similar in color and spectral characteristics, to have the same absolute brightness. By matching some of the stars of a cluster to similar stars for which the distance is known, we can derive the distance to the cluster. We can then determine the luminosity of the other stars in the cluster. Unfortunately most of the galactic clusters are obscured by interstellar material. This material not only absorbs light, but also reddens it, making the surface temperature of a star seem lower. By observing these stars in several colors, however, it is possible to derive intrinsic colors and surface temperatures.

With Sandage's period-luminosity-surface temperature relationship apparently well sustained, we can now determine the distance of a single cepheid if we know its surface temperature and period. . . .

The final result of these studies of cepheids in galactic clusters should be a useful and accurate period-luminosity-surface temperature chart for the cepheid variables. Astronomers may expect soon to have a much more reliable scale for measurement of long distances inside our own galaxy and beyond.

Philip H. Abelson*

ORGANIC GEOCHEMISTRY

From *Journal of the Washington Academy of Sciences,* May 1963.

Since life began on earth, living forms have synthesized a mass of organic chemicals about equivalent to the weight of the earth. Most of this material has been destroyed or consumed through biological action, but a significant fraction remains, mainly distributed in sedimentary rocks, which make up about 80 percent of the rocks of the surface of the continents.

Averaged over the world's entire surface, the thickness of these rocks is a little less than a mile. On the average, below each square centimeter of the earth's crust there are 1300 grams of organic chemicals or carbon of biological origin. These substances are found in rocks of all ages, including both recent and some of the oldest rocks on earth. Part of these chemicals are found as coal, oil, and natural gas, but the commercial occurrences represent a very small fraction of the total.

Moreover, in terms of scientific significance, the commercial aspects represent only a fraction of the interesting problems. Organic chemicals in rocks undergo with time and temperature a vast series of chemical transformations which challenge one's ingenuity to unravel. In the old sedimentary rocks are buried chemicals remaining from ancient life. These carry with them a potential wealth of information yet to be deciphered but nevertheless knowable.

* Dr. Abelson is represented by two articles in this volume. See also his "Conditions for Discovery" beginning on page 27 and the biographical note preceding it.

103

In an oxygen-containing environment, organic matter is usually speedily destroyed, although in a few notable instances, preservation of material has occurred under aerobic conditions. Under very dry conditions, organic matter may escape bacterial action for a long time; thus it has been possible to determine the blood types of some of the Egyptian mummies. Given long enough time, however, organic matter is degraded or destroyed in the presence of oxygen even without biological action.

A familiar example is the action of air on fats. Most living matter contains roughly a third of this material. In turn, most fats contain unsaturated fatty acids. The homemaker knows that fatty meats become rancid in the refrigerator or even in the deep freeze after a few months. Other materials are less easily affected by oxygen. Some amino acids are fairly resistant, but at room temperature in air even the stablest amino acids are destroyed in about 100,000 years.

The action of oxygen is often more rapid when the substance involved is colored. I recall isolating some porphyrin from an old rock where it had been preserved in an anaerobic environment and in the absence of light. The porphyrin had remained substantially unchanged for more than 400 million years. This same material, when extracted and left in solution in a flask in the light, was destroyed in 24 hours by the combination of oxygen and the sun's rays. Fortunately for geochemistry, there are many environments in which oxygen is absent, and under these circumstances material can be preserved. Anaerobic organisms take their toll, but they are much less efficient in destroying the organic matter than are the aerobic organisms.

The natural anaerobic environments usually are wet, and it is not surprising that water has a significant role in effecting chemical changes. Shortly after deposition, fats are hydrolyzed to glycerol plus free fatty acids; glycerol disappears, but the acids or their salts often remain. Complex carbohydrates tend to be broken down to simple sugars; these smaller soluble molecules tend to be lost. Rarely, under favorable circumstances, cellulose and chitin (polymerized acetyl glucosamine) may persist for tens of millions of years. As we shall see, proteins are hydrolyzed in a damp environment in about 50,000 years, but under special circumstances some of the amino acids may remain.

Perhaps the most important chemical events affecting the organic matter are reactions among the constituents themselves. The components of living matter are highly reactive. In the organized cell their mutual interactions are limited, but on death and lysis of the cell many reactions can occur. For instance, in an alkaline environment the aldehyde groups of liberated carbohydrates react readily with amines to make non-biological materials. Within a year or two after deposition of organic matter in an anaerobic environment, profound changes have occurred. Biochemists have a simple set of procedures to fractionate the components of living matter into carbohydrates, lipides, and proteins. The fats can be isolated by solvent extraction and the amino acids of proteins liberated through acid hydrolysis. The organic matter in sediments presents a puzzling problem. In a short period and at temperatures of 5° to 20°C., profound changes have occurred. Carbon, hydrogen, oxygen, and nitrogen analyses are only moderately different from the original living matter, yet solvent extraction allows one to isolate only a tenth of a percent or less of the original fatty acids. Similarly, although much bound nitrogen is present, only a small fraction of it can be isolated as amino acids.

The residual organic matter, often called kerogen, is practically insoluble in organic or inorganic solvents. It behaves like a plastic or a polymer of very high molecular weight. In an anaerobic environment, kerogen is apparently immune to biological attack. Being insoluble, it is not moved around by percolating ground waters. These properties lend great survival value, and it is not surprising that about 95 percent of the world's organic matter is in the form of kerogen. Asphaltenes, which are insoluble in water but can be extracted by carbon tetrachloride, constitute most of the remainder. However, a fraction of the original organic matter escapes incorporation into kerogen or asphaltenes and is found in rocks in an extractable form. These substances are present in low concentrations, but new techniques and modern instrumentation make them readily accessible for study.

In 1953 it occurred to me that organic matter could be preserved in shells and bones under conditions where it might be free from bacterial attack. This fortunately proved to be the case. When a calcium carbonate-secreting organism forms its shell, it employs an organic matrix. Thus, one can dissolve a clam shell in trichloroacetic acid and find thin sheets of protein. If one examines shells that have been buried for a few years, the accompanying protein is not perceptibly different from that associated with living clams. After a few thousand years the layers of protein in the shell are no longer as resilient as previously, but they contain the same amino acids as before and have been little hydrolyzed. After 50,000 years in a moist environment, the proteins have largely hydrolyzed and most of the amino acids are present in the form of small peptides or even as free amino acids. Through the action of ground water, part of these tend to be lost, but even after millions of years the shell still contains some of the original amino acids.

One of my favorite collecting spots is on Chesapeake Bay. The shells in the vicinity of Scientists Cliffs were deposited about 25 million years ago. Many of these contain about a hundredth of a percent of amino acids. The technique for isolation of these is quite simple. About 5 grams of shell are dissolved in hydrochloric acid. The resulting solution is made up to a volume of 100 ml and placed on a Dowex-50 ion-exchange resin which is in the H^+ form. On washing with distilled water, calcium and the amino acids are retained, while hydrochloric acid is removed. After washing with water the column is treated with six volumes of 5N ammonium hydroxide. Owing to the amphoteric nature of amino acids, they are eluted while the calcium is retained. The ammonia solution is taken to dryness, leaving a residue containing the amino acids and very few impurities. Components of the mixture can subsequently be identified by paper or column chromatography.

Using these procedures I have examined a large number of shells, bones, and sedimentary rocks and have found amino acids in environments that were deposited as much as 450 million years ago.

There are some limitations. All amino acids are not sufficiently stable to exist for so long a period of time. Serine, threonine, and arginine can remain for only about 10 million years. Amino acids have been found only in geologic settings which have been clearly anaerobic. Amino acids have not been found in fossils showing evidence of recrystallization. Even the most stable amino acids are destroyed if exposed to elevated temperatures. Laboratory studies show that alanine could be expected to persist after three billion years at 25° C., but that at 100° C. it would last about 100,000 years. In the older fossils I found alanine, glycine, glutamic acid, the leucines, and valine. These are the most stable of the group. The laboratory studies show that the 4-carbon *a*-amino butyric acid is as stable as alanine or valine, and there have been synthesized a large number of amino acids which would be equally stable. In my studies of old rocks I did not observe any of these amino acids. I was able to show that the organisms of 300 million years ago were using some of the same amino acids as are used today. If they had been employing sizable quantities of some of the other stable entities, it would have been possible to find them. The absence of these other substances is evidence for an unchanging pattern of utilization of amino acids. . . .

*O*ne of the great researchers and interpreters of American pre-
history is Dr. Alfred V. Kidder, former chairman of the Division of Historical Research of Carnegie Institution. His "Excavations at Kaminaljuyu, Guatemala," published in American
Antiquity in October 1945 gives a fascinating picture of the problems and experiences of a field explorer in archaeology, and
discloses some extremely interesting information about the peoples
who once lived in the vicinity of modern Guatemala City.

"Excavations at Kaminaljuyu, Guatemala," is one of a
series of readable and significant papers and books written by
this distinguished explorer of America's pre-Columbian past. Dr.
Kidder's publications include "A Program for Maya Research,"
The Pottery of Pecos, The Artifacts of Uaxactun, Guatemala,
"Looking Backward" and many others.

Dr. Kidder was born at Marquette, Michigan, in 1885.
He received his A.B., M.A., and in 1914, his Ph.D. degree, from
Harvard University. He took to archaeological research early; at
the age of 22 he was already engaged in research explorations of
Colorado and New Mexico for the Archaeological Institute of
America. The following year he carried on explorations in Utah
for the University of Utah and in New Mexico for Harvard University. In 1909 he traveled in Egypt and Greece. In 1910, and
again in 1912 and 1914, he served as Austin Teaching Fellow at
Harvard, meanwhile continuing research explorations in New
Mexico and Utah for the New Mexico Territorial Museum and
other institutions. He was Curator for North American Archaeology of the Peabody Museum, Harvard, in 1914, and directed excavations at Pecos, New Mexico, for Phillips Academy, Andover,
Massachusetts, from 1915 until 1929.

He was appointed Associate of the Carnegie Institution
in charge of archaeological investigations in 1927, and became
Chairman of Carnegie Institution's Division of Historical Research
in 1930, in which position he served until his retirement in 1950.

108

Alfred V. Kidder

EXCAVATIONS AT KAMINALJUYU, GUATEMALA

From *American Antiquity,* October 1945.

I suppose every digger has at one time or another yearned for a potoscope or dreamed of an archaeologically endowed hazel bush whose twigs would tug and dip to well-stocked graves. But if such sure-fire indicators existed, I doubt if fieldwork would be as much fun as it is. Would fishing have quite the same lure if every cast brought a strike? For better or worse, archaeology is a great gamble; each new site offers unlimited possibilities of surprise. So although the introductions to our reports often let it be inferred that our finds have been the result of keen scientific foresight, very few excavations, I imagine, turn out as we expect them to. It was certainly that way at Kaminaljuyu.

The word Kaminaljuyu is quite a mouthful. It means, in the Quiché language, "Hills of the Dead" — the *juyu* being the hilly part of the term. Which reminds me of S. K. Lothrop's remark that the name of another Guatemalan site, Cheerijuyu, would make a good toast at an archaeologist's party. Kaminaljuyu, or K.J., as it is known to its intimates, rates a long name, for it consists of some two hundred mounds covering an area of about five square kilometers on the outskirts of Guatemala City. It has been known for a long time, having been mentioned as early as the eighteenth century by Fuentes y Guzman. A hundred years elapsed before it again came to notice and then it was contemptuously dismissed by C. H. Berendt, who said in the Smithsonian report for 1877, "There might be named hundreds of other places in Central America more likely to give returns." No excavation was done there until 1925, when Manuel Gamio, the Mexican archaeologist, made stratigraphic tests on the Finca Miraflores, one of the several farms into which the site is divided. Gamio's work had the important result of bringing to light potsherds and clay figurines recognized by him as analogous to those of the Valley of Mexico Archaic. Kaminaljuyu was thus shown to have been settled at an early date. Shortly thereafter, Lothrop published a study of its numerous stone sculptures, which indicated that its occupancy had extended well into the Classic or Old Empire period of the Maya.

Gamio's finds and Lothrop's data on the sculptures made it evident that Kaminaljuyu had had a long pre-Columbian history. It therefore seemed to be a likely place for finding deposits representative of at least two phases of Guatemalan highland culture. So, in 1935, Oliver Ricketson and I put in a couple of weeks doing another test section at the place where Gamio had worked. But although we got a large amount of material, we had the bad luck to pick a spot in which the ancients had dug several deep pits. These had messed up the stratigraphy pretty badly. We therefore planned, for the next season, to look for a less disturbed part of the midden. But . . . it was three years before we had time even to look at Miraflores again. What happened was this:

The then Secretary of Education, Dr. J. Antonio Villacorta, eminent historian and, when opportunity offered, excavator, had some years before dug in one of the Kaminaljuyu mounds and uncovered a stairway of burned adobe. He had been, perhaps, a little over-enthusiastic about this discovery — or more likely the newspapers played it up too strongly. At all events, unkind persons held that the Secretary had found nothing more exciting than an old brick kiln. There was, I have heard, quite a controversy. So Dr. Villacorta naturally kept his eye open for any further evidence which might prove his stairway to have been the real McCoy. His opportunity came in the autumn of 1935.

On the Finca La Esperanza, at the eastern edge of the site, were two grass-covered mounds, fairly good-sized but far smaller than scores of others at Kaminaljuyu. No one had paid any attention to them and probably never would have if they had been 100, rather than 90, meters apart. That 10-meter deficit made the flat ground between them just too short for a regulation association football field, so a local club that wanted to use the land got permission from the owner to lengthen it by cutting a slice off the face of each mound. The work on one of them had been done and was under way at the second when Dr. Villacorta happened to drive by, saw what was going on, and went in to have a look. He found that a sloping, plaster-covered surface was being destroyed, stopped the digging, and had his son Carlos, then Director of the National Museum, run in a small exploratory trench. This showed that the slope formed part of a buried structure, apparently in good condition. The trench was then refilled and Dr. Villacorta asked if the Carnegie Institution would care to make further investigations. We agreed to do so, for the potsherds found by Carlos Villacorta were not of Miraflores types. Furthermore, aside from the "brick-kiln" stairway, no remains of buildings had appeared at any of the Kaminaljuyu mounds and we were keen to learn something of the site's architecture.

Oliver Ricketson and I accordingly went out and examined the mound. It was about six meters high and twenty in diameter. The surface exposed by the footballers seemed to be the basal slope of some kind of substructure which we figured could be cleared in three weeks or so at a cost of about a hundred dollars. We started work a few days later, but before a month was out I was cabling Washington for a special extra appropriation. Those three weeks stretched into three full field-seasons, two more years in repair and study of the specimens we found, and a longer time than I like to think of in getting the report written. . . .

Mound A, the one in which the sloping surface had appeared, turned out to be like one of those trick Chinese wooden eggs, in that it contained no less than eight structures, one inside the other. In addition, there were six pit-tombs, evidently the graves of important and wealthy personages (Fig. 1*a, b*). Mound B was somewhat, but not very much, easier, for it had only five structures and four main tombs. . . .

We started well outside the mound with a series of trenches sunk to and run in along the top of sterile subsoil. Before we even reached the original exposure we got the first of the long succession of surprises that those two mounds had in store for us. We had supposed Mound A to be a simple, one-period structure and that the sloping surface exposed by Carlos Villacorta had been covered by debris from the disintegration of its upper parts. But outside it we encountered the remains of a later building. This was so badly gone — we afterward found it in better shape on another face of the mound — that we noted what we could, calling it, in our innocence, Structure B. We then cleared it away to get at the more promising Structure A. The lower parts of A proved to be in fine condition. It was pyramidal and had been built of waterworn lumps of pumice laid in stiff black clay. This hearting was veneered with an almost concrete-hard mixture of lime and volcanic gravel, finished with pure white lime. There was a basal step from which rose a sloping zone (the element first encountered); the latter was topped by a molding-framed vertical zone. These constituted the first terrace, a second was of identical construction and carried the pyramid to a summit platform that had borne a temple which had been completely razed by the builders of Structure B; the ground plan of its single chamber could be determined from the holes that had held its four corner-posts. The restoration of the superstructure drawn by

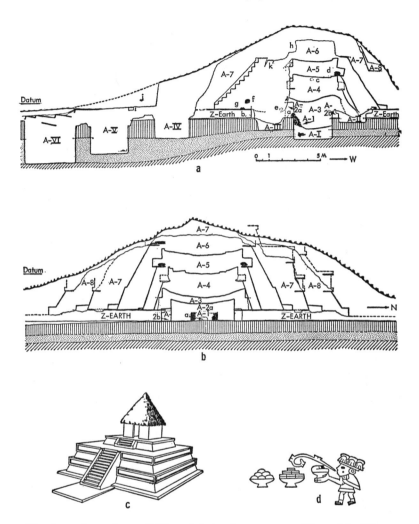

Fig. 1. *a, b,* sections of Mound A to show superposition of buildings and relationship of buildings to tombs. In the east–west section, *a,* most of the lower-case letters indicate caches of ceremonial pottery; *f* marks the position of a jade boulder; *b,* that of the skeleton of an apparently sacrificed young woman; *c,* restoration by Tatiana Proskouriakoff of Structure A-7. The only doubtful element is the thatched temple roof, so restored because no stones suitable for a Maya-type vault were found. *d,* priest presenting offerings.

Miss Proskouriakoff (Fig. 1*c*) is, aside from the basal dimensions, entirely hypothetical. The substructure, however, is correctly rendered, although we were in doubt as to details above the lowermost sloping zone until the second season; and not certain regarding all of them until the third year, when we excavated Mound B and found parts of a practically identical building in better condition.

Structure A (it afterward became Number 7) at first puzzled us greatly because it appeared to have no stairway. But when we had cleared the whole front we found a gap of about the right width in the sloping lower zone. We guessed — and this was one of the few times we were correct at the first go-off — that this marked the former position of the stairway; and as we wanted to examine the pyramid's fill, we took advantage of the gap to push a cut inward. And immediately we got our second surprise, for an hour's digging brought to light a flight of adobe steps in perfect preservation. So now we had three structures and, clinging stubbornly to our original nomenclature, we called this one A-sub.

To make a long story short, we found five more: the earliest three rectangular, vertical-walled blocks of adobe, which probably did not have superstructures; the next three adobe pyramids, which certainly did, as we found postholes and, in one case, the clear print of the wall; finally, two pumice and mud pyramids. The degree of exactitude with which we were able to determine the original form and dimensions of the successive increments naturally varied, but in every case it was possible to recover reasonably satisfactory data. We were amazingly fortunate. Of one pyramid, for example, the stairway was perfectly preserved, but above it almost nothing had survived. Of a second, we found only vestiges of the stair, but the entire summit platform was intact. In still another case, a narrow strip of the platform remained to indicate its former height. Furthermore, several of the structures had obviously been so much alike that the nature of elements missing from one could often confidently be postulated from corresponding parts of another. We made a number of blunders, some relatively serious, some apparently unimportant. In extenuation of our shortcomings, I may say that as we were dealing with remains of a type entirely unfamiliar to us; we never knew what we were likely to encounter. Our approach, therefore, though cautious, was often fumbling. In addition, each of the seven inner buildings had been considerably damaged, and certain of them had been almost completely destroyed in the course of the erection of their successors. Recent terracing and the cutting involved in the lengthening of the football field had removed other important evidence. Finally the very extensive settling which took place as a result of the failure of tomb roofs or the compacting of their fills caused most confusing vertical displacements.

Even the lucky accidents of preservation just mentioned would not have served us as they did, had it not been for the intelligence, keen observation, and limitless patience of my two assistants, J. D. Jennings and E. M. Shook. Jennings, who was with me the second year, did the really difficult work on the remnants of the five innermost structures in Mound A, while I gave myself the much more congenial task of clearing tombs. And Shook, in the third season, handled the equally difficult job of dissecting Mound B, I, in the meantime, working on potsherds. No small part of Jennings' and Shook's success was due to their careful coaching of the workmen, who quickly became interested and who proved to be better than any of us at spotting the subtle differences in texture and color which often gave us our first indication that we were passing from one adobe building into another. On several occasions they saved us from cutting away construction the loss of which, before it was recorded, might have been disastrous.

To get back to the first year's digging, I had noticed, before the central approach trench had quite reached the mound, a yellow area in the subsoil. At the time, I put it down as a natural outcrop of a volcanic deposit locally called *talpetate,* which appears on the surface here and there at Kaminaljuyu. But thinking it over, I got a little suspicious and had the trench floor recleaned. Close examination showed flecks of charcoal in the *talpetate,* and as these had no business to be in an undisturbed formation, I went deeper. Soft, dark earth full of potsherds appeared. I put a man to sinking a test-hole. He went down and still down. This was exciting. It looked as if we had hit a tomb but I was awfully afraid it might be only a rubbish pit. At three meters from the surface, though, we came to a floor; and right where we struck it there lay a crushed pottery vessel and fragments of bone!

I should, of course, have laid off the tomb until the general situation was clearer (at that time I was running the work alone, for Ricketson had been taken sick), but I was so anxious to see what it contained that I at once had the pit enlarged, thereby destroying evidence I later needed very badly as to its fill and as to the grave's relation to a platform that was afterward found to have covered it. Nor did I handle any too well the cleaning and recovery of the material it contained. Fortunately, however, this tomb, although it held many fine things, was the poorest at Mound A and was much the best of the series to have cut my teeth on. I got from it much valuable information as to how interments of this sort should, or rather should not, be excavated.

While I was working in the tomb, I kept thinking about the ripped-out stairway of what I was still calling Structure A. The digging at that point had not yet reached bottom and I wondered if the stairs could have been removed to put in another tomb. I therefore sank a pit in the gap and sure enough, there *was* a tomb. This and the first one were aligned on the central axis of the mound. Maybe, I thought, there might be still another further out, beyond the beginning of the approach trench. There was. I certainly had a bull calf by the tail. So, exercising almost superhuman restraint, I devoted myself to work on the mound proper until Robert Wauchope could finish an excavation he was doing at Zacualpa in the western highlands and come in to help me. When he arrived, he and Mrs. Kidder and I tackled the two new tombs, finishing them just as the onset of the rainy season put a stop to that year's digging.

The second season saw Mound A completed. We found three more tombs there as well as a line of three directly in front of Mound B. The third year was put in on Mound B. It proved to be much like A. There were three inner structures of adobe and two outer ones of clay and water-worn pumice lumps faced with the same hard mixture of lime and gravel. Parts of the inner one of the latter pair were in fine shape, providing us with much useful information as to the original appearance of this latest and finest type of building. Underneath the mound were additional tombs.

Of the twelve tombs, all but a single minor one at B were located on the central axis of the mounds. The putting in of each evidently coincided with the dismantling of a structure and the erection over it of a new and bigger building. Such a correlation of burials with large-scale architectural activity suggests that the individuals interred must have been persons of much importance. This is further borne out by the size of their graves and the richness of their mortuary equipment.

The typical tomb was a rectangular, straight-sided pit three to over four meters long by slightly less wide, and often three to four meters deep. In the case of the two earliest at Mound A, the bodies had been at length and later burials had disturbed earlier ones. But all the ten others had been used only once and funerary practice had followed a remarkably uniform pattern. The principal occupant of the tomb had been placed in the middle of the floor seated cross-legged, facing south, hands in lap (Fig. 2). He—for all were adult males—had been coffined in a capacious container, apparently a

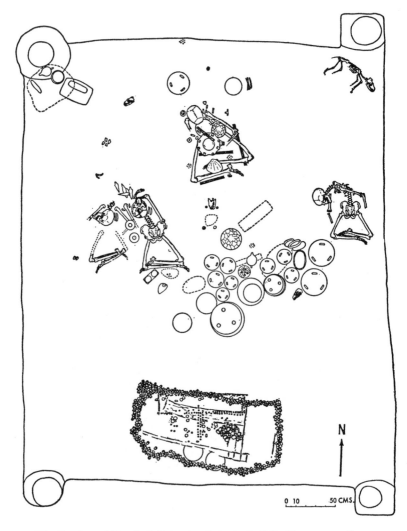

Fig. 2. Plan of Tomb A-IV, a tomb typical of the Esperanza phase.

wooden box. The corpse had been literally loaded with jewelry; jade necklaces, earplugs, pendants, shell ornaments, pyrite-mosaic "mirrors"—each of two tombs held no less than six of these beautiful and costly objects. Between the body and the south wall there was always a pile of fine pottery vessels, many of them brought from great distances. Other vessels, "mirrors," conch-shell trumpets, obsidian spearpoints, alabaster cups, lay here and there on the floor. Every tomb but one contained a small three-legged metate with its grinding stone; in one there was a flask of liquid mercury. Such were the imperishable objects, but these had formed only a part of the lavish offerings. There had been bundles of textiles, reduced by decay to areas of soft, fluffy black rot; gourd or wooden trays and containers stuccoed and elaborately painted; also what seemed to have been masks or headdresses of carved wood, stuccoed and painted and embellished with discs of mica, the mouths perhaps set with jaguar jaws, of which we found a number of pairs. In two of the largest tombs could be made out the remains of pole-and-rod litters or biers, one of them inlaid with pyrite and hematite discs and covered by a pall fringed with hundreds of *Oliva* shell tinklers.

These ancient tycoons had not been allowed to go unattended into the hereafter. In all but the one smallest tomb were the remains of two, or three, in one case four, other persons. As well as we could make out—all skeletons were in the last stages of decay—most of these were young females, probably slaves or concubines sacrificed at the time of their master's death. They had few or no ornaments and had been buried, also seated cross-legged, near the walls. In two of the pits the skulls of children and adults replaced the youthful skeletons; in another, there were both skeletons and an elaborately carved adult skull; in several were the skeletons of small dogs. The tombs had sometimes been filled at the time of burial, sometimes left open. But in either case they had been roofed with heavy logs. The collapse of the roofs and the impacting of the fill had caused much downward displacement of overlying structures.

The clearing of those pits was a fascinating job, but a mortal hard one. They contained so many and such beautiful objects and almost everything crushable had suffered so cruelly from the enormous pressure of slumped buildings and platforms, that the utmost care had to be exercised. Most of the work was done with knives and fine brushes; from some of the stuccoed pieces the earth could be removed only by whiffing it gradually away with a rubber air-bulb as it dried. Kneeling or squatting in cramped positions day after day—no tomb took less than a week to clean and one required nearly two—would have been bad enough in the open; at the bottom of those holes, with all breeze cut off and the sun beating relentlessly down, it was at times pretty close to purgatory.

In doing the first tomb, I made the mistake of coming down from above, thereby damaging some specimens before it became evident what they were and how they should be handled. In the next two, a small pit was sunk in one corner to locate the floor. Then the fill was removed to within 40 cm. or so of that level and we worked out laterally along the floor. That was better, but the vertical excavation of the fill prevented making accurate observations on remains of roofing, collapsed platforms, and debris from dismantled buildings, all of which could yield important evidence for the history of the tomb.

In the procedure finally adopted, excavation was entirely from the side. The size of the grave was first determined, a simple matter as the dark, relatively soft fill could easily be distinguished from the clean, hard clay through which the shaft had been sunk. Then a pit as wide as the shaft and sufficiently large to give ample room for shovelling was dug just outside the tomb on the north, that side being chosen because we had learned that mortuary furniture seldom lay close to the north wall. The pit was carried down to the level of the tomb floor, exposing the entire north face of the fill. A diagram recording its composition and stratification was made, after which the work was pushed southward, the floor of the excavation being kept at about 50 cm. above that of the tomb. The fill was taken out in meter-square columns. Columns, rather than tomb-wide slices, were cut because the latter would have permitted observation of only east-west faces, whereas diagrams of the two faces of each column could later be combined to form a series of both east-west and north-south sections at 1 m. intervals. The bulk of the fill out of the way, the 50-cm. layer left on the floor was taken out, also in blocks 1 m. square. This final operation was of course done entirely by hand. Work began at the north end of the tomb and was carried southward, specimens and bones being brought to light, noted, recorded on the ground-plan, and taken up.

At first we attempted to clear the skeletons and vessels in order to get photographs of the entire contents of a tomb. But we soon found that materials left in situ prevented access to others, that some of them deteriorated badly during the several days they often had to lie open to sun and air, and that they were in danger of being stepped on or damaged by falls of earth from the drying walls of the tomb. Furthermore, decayed bones and shell, rotted "mirrors," and crushed pottery, particularly vessels coated with stucco, suffered severely from the brushing necessary to expose them for photography. Hence we removed each specimen as we came to it, the more fragile ones, as soon as their nature and size were ascertained, being undercut and taken out, still encased in earth, for final treatment in the laboratory. In such cases, exposed parts of objects were covered with several layers of moistened tissue paper and the whole lump was solidified with a heavy coat of paraffin.

We were, of course, very fortunate in being so handy to our work-rooms in the city, where the lumps could be dissected at leisure; and even more fortunate in that Dr. Wissler of the American Museum of Natural History and Mr. Scott of the Peabody Museum of Harvard allowed Mr. Paul Richard and Mrs. Harriet S. Cosgrove of their respective staffs to come to Guatemala to do this delicate work and to mend the specimens. Their efforts were effectively supplemented by Sr. Cesar Tejeda, who restored much of the pottery and Sr. Antonio Tejeda, who drew or painted the vessels and other objects.

The most difficult material to handle was the stuccoed

pottery, of which some tombs contained a great deal. The vessels were not only crushed, but the paper-thin stucco, embellished with most beautiful paintings, had in many cases peeled off in hundreds of small flakes that tended to curl on drying. They seldom corresponded in size or shape to the fragments of the vessel; often three or four vessels had been mashed together. It was double-barrelled super-jigsaw. The pyrite-incrusted stone plaques, usually, but I think probably incorrectly, called mirrors, were almost equally hard. The pyrite mosaic which covered one side of each had completely disintegrated, but the other side had often borne painted stucco and in one case was elaborately carved. In trying to get it out of the earth, I completely ruined the first one I came across, not realizing that it was riddled with tiny cracks. After that all "mirrors" were removed *en bloc* and cleaned in the laboratory.

About the only things that gave us no trouble were the jades. The quantities of this, to the ancient Mesoamericans most precious of all substances, indicated the importance of the men whose corpses it bedecked. The examples in the Kaminaljuyu tombs were not as beautifully carved as were those of later periods, but the stone was of superlative quality, some bright emerald, some a clear very dark shade. I know of no greater thrill than to have one's carefully probing knife-point skid on a jade's polished surface—you can't mistake that for the feel of anything else—and, shifting to the brush, see the first green gleam in the brown earth.

The tomb jades were beads, pendants, earplugs, mosaics, the thin plate-like elements of the latter unfortunately detached from their backings. In addition to these ornaments, there were, in two of the tombs, collections of jade-worker's materials: sawed pebbles, drill-cores, beads in various stages of manufacture, a "blank" from which an earplug was to have been fashioned. And buried as a dedicatory cache in the center of the adobe stairway of Structure A-6 was a waterworn boulder 40 cm. in diameter into which a few exploratory cuts had been made. This, I believe, is the largest piece of American jade ever found.

During such an excavation as that of the Kaminaljuyu mounds and tombs one is so hard put to keep up with the delicate hand-work that must be done by the staff and with the routine mapping and photographing and note-taking, one is so preoccupied with the immediate problems of what next to do and how best to go at it that there is little time to think of the wider implications of what one is finding. We knew from the first that we were getting useful data. As a matter of fact, any information from the Guatemala highlands was bound to be valuable because that region is archaeologically next to unknown. But only when the specimens had been studied and the field-records reviewed could we see our results in proper perspective.

We found that we had some data—mostly on ceramics and figurines—regarding the early Miraflores phase first brought to light by Gamio on the finca of that name. The Miraflores remains are analogous to, and probably roughly contemporaneous with, those of the so-called "Archaic" cultures of Mexico, of the early Mamom and Chicanel phases of the lowland Maya country, the pre-classic phases of Honduras and of Salvador. We also found evidence that the site had still been occupied in late prehistoric times. However, the bulk of the material derived from the two mounds and their tombs is of an intermediate phase which we have called Esperanza.

The Esperanzans without much question spoke a Maya tongue but their culture, although highly ceremonialized, differed from that of the classic lowland Maya as exemplified by Tikal, Uaxactun, Palenque, and Copan, in that corbelled vaults were not built, nor were dated stelae erected; and the run of the pottery was of types not found at the classic Maya cities. I say the run of the pottery because the tombs also contained certain typical Maya vessels, evidently trade-pieces, which show that the Esperanza phase was contemporaneous with the early part of the Old Empire and therefore is datable—if the Goodman-Martinez-Thompson correlation of Maya and Christian chronology is correct—at about the sixth century A.D. Many of the vessels had been imported from Teotihuacan or some other center of that then flourishing Mexican culture; and the architecture of the two latest buildings in each of our mounds corresponds, detail for detail, so closely to that of the great period at Teotihuacan that it seems possible that an invasion from the north may actually have put Mexican priest-rulers into the saddle at Kaminaljuyu. Other pieces of pottery show influence from the great Oaxacan metropolis at Monte Alban during its third period; others are from Salvador; still others—of a beautiful creamy ware—are from some as yet unidentified center. Finally, the last tomb we excavated yielded a carved "mirror" in the unmistakable Totonac style that reached its height at Tajin in Veracruz.

Thus the Kaminaljuyu finds make it clear that during Esperanza times an active exchange of commodities was going on all over Mesoamerica. The above-mentioned specimens are very valuable archaeologically, for they serve to tie together in time several of the most important Mesoamerican cultures. They also suggest that there was an almost simultaneous burst of cultural activity throughout the area. Hence neither the Maya nor the people of Teotihuacan—both of whom have had their advocates—can be considered to have been uniquely outstanding as sparkplugs, so to speak, of the classic Mesoamerican development. This throws one back, in the attempt to reconstruct the growth of higher Mesoamerican culture, into such earlier phases as Miraflores, Mamom and Chicanel of Peten, the Mexican Archaic, etc., which, although certainly less advanced than the classic cultures, are by no means primitive and, as we come to know them better, are proving to have attained stronger local specialization and greater ceremonial development than was once believed. Such being the case, one has to look still further into the past for ultimate beginnings. And there one faces a blank wall, for nothing that can be recognized as ancestral has yet come to light. That, at present, is the great enigma of Mesoamerican archaeology.

I have wandered a long way from Kaminaljuyu and from the Esperanza phase, the only period in the life of that site with which we are even reasonably familiar. We know little enough. Nevertheless, the two mounds, so often enlarged at such great cost of labor, and the dozen tombs so richly stocked with costly offerings, served to indicate an economy sufficiently thriving to free many people for occupations above and beyond the mere getting of a living. Government was presumably theocratic and, to judge from the extent of public works, highly autocratic. It doubtless centered at Kaminaljuyu, which was evidently the administrative center for a large region over which were scattered the farming villages of the commoners. From these, people came to the city for religious festivals, for the market days that are still so great a feature of Guatemalan life, and, in seasons when their crops did not demand attention, for work in the constantly growing assemblage of temple-crowned pyramids. Trade was brisk. Seashells, tropical fruits, and cotton came from the nearby Pacific coast plain; brilliant quetzal feathers from the mountains. For export the Esperanzans had abundant supplies of obsidian, an invaluable stone for cutting-tools much in demand in the thickly populated non-volcanic lowlands of the north. I also suspect that they had some nearby source of jade. Whether the pottery from the Maya country and that from even more distant regions in Mexico was brought by organized trading parties or was passed from tribe to tribe, is unknown. Long-range commerce, among the American Indians, seems to have gone on more or less uninterruptedly in spite of wars. Conditions in highland Guatemala at this period seem, however, to have been relatively peaceful, for Kaminaljuyu and apparently all other large centers of Esperanza times occupied open sites, quite unlike the barranca-guarded strongholds which existed in later centuries.

The rulers, as I have said, may have been Mexicans. Whoever they were they topped a governmental organization heavily weighted on the ceremonial side. How many of the more than two hundred mounds at Kaminaljuyu are of the Esperanza phase is not known; I think a large proportion. The religious cults of that period, to judge from the paintings on the stuccoed vessels from the tombs, contained both Maya and Teotihuacan elements. In one respect Kaminaljuyu is unique. It contains more ball-courts than any site so far surveyed, at least thirteen, and there are many more at other mound groups in the valley. It must have been an even better ball town than Brooklyn.

The ball-game was doubtless to a large extent a ritualistic observance. The other mounds were, of course, purely religious structures. The tombs we found were without much question those of theocrats. Almost all our information so far—and this is true throughout Mesoamerica—has to do with the ceremonial setup. About this we have learned a great deal. But as to the people who labored under so heavy a weight of service to their rulers and their rulers' gods, we know very little. Some day we shall get around to excavating the village sites and the little unspectacular house mounds. These will give us an insight into the domestic arrangements of the commoners: their houses, their implements, their methods of cooking. Their graves may teach us, perhaps, something of humble folk-cults. Such work will also be of the greatest archaeological significance, for the small amount of digging we did outside the mounds showed that temple furniture and the mortuary equipment of the overlords was in a class by itself. It no more represented the general contemporary culture of Esperanza Kaminaljuyu than the contents of a Park Avenue drawing room does that of the United States. Only when we have much fuller knowledge of everyday household utensils, corn-grinders, common stone and bone tools, objects which were not subject to shifting whims of fashion or influenced by new and perhaps quickly and widely spreading cults, shall we be able to discern the basic relations of cultures.

This is but one of the manifold tasks that face Americanists in Mexico and Central America, in the Eastern United States, in the Southwest, in the Andes. But they are well worth the doing, for in these continents we have a unique opportunity to study the rise and the spread of what, in spite of the fact that it was overwhelmed by the white man's coming, was one of the world's great civilizations. If we do our work as we should, we may come to understand the meaning of that civilization's strange pulsations, we may learn what stimulated its successes and what brought about its failures. We may even glean some knowledge—and the Lord knows we need it—of the mysterious forces that in all ages and in all parts of the globe seem inevitably to have caused each hopefully flowing tide of human progress to halt and finally to ebb.

*A*s head of Carnegie Institution's Department of Embryology, at Baltimore, Maryland, Dr. James D. Ebert directs one of the world's foremost research institutions devoted to further understanding of human and animal development in the vital and little-known period from conception to birth.

An important part of these researches involves the processes by which the original cells of an embryo differentiate and develop into specialized organs and other structures. Since all of the cells in a given animal presumably carry the same genetic information, it is still a mystery how some of them turn out to form liver, others kidney, and the like. One of the most interesting organs for this kind of study is the heart, for not only is the heart commonly identified with life itself, but it is also one of the first specialized organs to appear. Much has been learned about the heart's early history at the Department of Embryology, and a leader in this research has been Dr. Ebert himself.

Dr. James David Ebert was born at Bentleyville, Pennsylvania, in 1921. After graduating from Washington and Jefferson College, Pennsylvania, in 1942, he postponed his scientific career to join the Navy. He served four years of active duty as communications officer, and later as gunnery officer, aboard the destroyer Dickerson until she was sunk in 1945. For his part in that action he received the Purple Heart.

Returning to civilian life, Ebert began graduate work in biology at the Johns Hopkins University in 1946. After receiving his Ph.D. degree in 1950, he devoted five years to research and teaching in experimental biology, first at Massachusetts Institute of Technology, and then at Indiana University. During the summers of 1953 and 1954 he carried on research at the Brookhaven National Laboratory, Long Island, New York.

On January 1, 1956, Dr. Ebert became Director of the Carnegie Institution's Department of Embryology, succeeding Dr. George Corner, who had retired. At the Department he has continued his investigations in developmental biology, especially his work on the early history and development of the heart. Among his other research interests are immunochemistry, which involves problems of tissue transplantation, studies of the role of amino acid and vitamin metabolism in embryonic development, and the effects of viruses on developing tissues.

128

James D. Ebert

THE FIRST HEARTBEATS

From *Scientific American,* March 1959.

In superstition, legend and sentiment the beating of the heart is synonymous with life itself. The last heartbeat surely marks the end of life. The first heartbeat does not, however, denote the beginning. Although the heart begins to beat when most of the other organs are still unformed and the heart itself is but a simple tube, embryologists have traced the origin of the heart to even more primitive stages of development. In this work the microsurgery of classical embryology is now extended by the powerful techniques of biochemistry. A battery of subtle chemical reactions has disclosed the formation of the first fibrils that go to make up the heart-muscle fibers, and has detected the synthesis of the contractile proteins that compose the fibrils.

In the human embryo the heart and major blood vessels develop almost entirely between the third and eighth weeks of life. By the second month, when the embryo is just over an inch long, the heart has assumed its adult form. The major sequence of events in the formation of the heart is essentially the same in all vertebrate animals. In this article the heart of the chick will serve to illustrate the process for the entire vertebrate order, from fish to man.

Two sets of primitive structures on opposite sides of the embryo give rise to the simple tube in which the first heartbeats occur. Each set consists of a delicate thin-walled tube and an adjacent ribbon of cells which are destined to become muscle. Near the mid-line of the embryo the tubes fuse to

129

form a larger tube sheathed with muscle cells. The fusion progresses from the head toward the tail, first forming the ventricle (which gives rise to the two chambers that pump blood out of the heart) and then the atrium (which gives rise to the chambers that take blood into the heart).

As the tube forms, so it begins to beat. Florence R. Sabin of the Johns Hopkins School of Medicine described the first beats in 1920 in a famous paper on the origin of the heart of the chick embryo; other investigators, using motion pictures and time-lapse photography, have added to the story. The first twitching can be seen early in the second day of development. The slow but rhythmical beat begins along the right side of the ventricle and gradually involves the whole ventricular wall, spreading from the tail end of the embryo toward its head. Soon the entire muscle of the ventricle is contracting synchronously, periods of pulsation alternating with periods of rest. Meanwhile the atrium has been forming below the ventricle. As the atrium takes shape, it too begins to contract, and at a more rapid rate than the ventricle. The ventricle, however, increases its rate of contraction to keep pace. The contractions now set the blood in motion.

The last region to develop is the "pacemaker," which controls the contractions of the fully formed heart. When this region starts contracting, the beat of the whole heart accelerates further. In the embryonic heart the region with the highest rate of contraction sets the pace for the entire organ; if the various regions are cut apart and isolated from one another, each tends to revert to its characteristic rhythm. What synchronizes their contractions at this early stage is a mystery; nerve fibers have not yet grown out from the central nervous system to the heart, nor has the heart's own internal system of communication been established. Perhaps the answer lies in the intrinsic contractile properties of the muscle, or in the activity of chemical regulators.

On the third day of development the ventricle, which is now U-shaped, assumes its ultimate position behind the atrium. The structures that divide the ventricle and atrium each into two chambers appear in the course of the fourth day. By the fifth day the chick heart is substantially complete.

So much is visible to the eye. But what events presage the formation of the primitive tube?

The chick embryo begins to develop while the fertilized egg is moving down the oviduct of the hen. In a newly laid egg the embryo is already large enough to be seen with the naked eye; it is a tiny white disk on the surface of the yolk. At this point it is composed of only two layers of tissue: the epiblast (including both the ectoderm and the future mesoderm) and the endoderm. If it is now cut up so that its fragments may be grown in tissue culture, pulsating tissue will grow from pieces taken from the edges of the embryonic disk but not from those taken from its center. Apparently the heart-forming cells are distributed around the periphery of the embryo at the end of the two-layer phase. As development proceeds, some of the cells of the upper layer migrate toward the mid-line of the embryo. There an elongated opening—the "primitive streak" —has formed. The migratory cells move through this opening into the interior to form the third embryonic cell layer: the mesoderm. Tissue-culture studies at successive stages in this process indicate that the heart-forming cells are among the migratory mesodermal cells.

Maps of the embryo based on such studies show that the heart-forming cells move first toward the tail end of the embryo and then through the primitive streak into the mesoderm, assembling in regions on either side of the head end of the primitive streak. Under the artificial conditions of tissue culture, or of transplantation to another site in the embryo, it appears that a much larger population of cells is able to form heart than actually takes part in building the organ. There is no striking difference of shape or extent between the right and left heart-forming areas, but the left side has a greater capacity to form heart muscle. This example of bilateral asymmetry in the embryo was discovered in 1943 by Mary E. Rawles of Johns Hopkins University, who mapped the heart-forming capacities of the chick embryo at this early stage.

In normal development the two heart-forming regions gradually move toward each other, and the primitive structures that arise in them merge near the mid-line. But one can prevent the two regions from joining by removing the wedge of tissue lying between them, by inserting a barrier or merely by holding them apart. The embryo will then develop two separate hearts. Both hearts usually have normal shape and orientation, but occasionally the right heart develops as the mirror image of the normal left heart.

Alteration of the chemical environment of the embryo can also induce double hearts to form. Recently my associate Robert DeHaan has found that substances such as "Versene" or acetylcholine, which are capable of binding or displacing calcium, exert this effect. Calcium and magnesium salts act as a sort of cement in the tissue structure; for example, calcium ions may interact with charged groups of atoms on the surfaces of adjacent cells and thereby hold the cells together. The calcium-binding agents make calcium ions unavailable at a critical period, disturbing the intercellular relations in the embryo sufficiently to prevent the movement of the heart-forming areas.

Although heart muscle arises from the mesoderm, we cannot say whether the process involves the mesoderm alone or the interaction of the mesoderm with other tissue layers. Many significant embryonic events involve such interaction. Experiments by Robert L. Bacon, then at Yale University, have indicated that interaction of the mesoderm and the endoderm is necessary for the normal development of the salamander heart. The situation in the chick is not altogether clear. The fragments transplanted for tissue-culture experi-

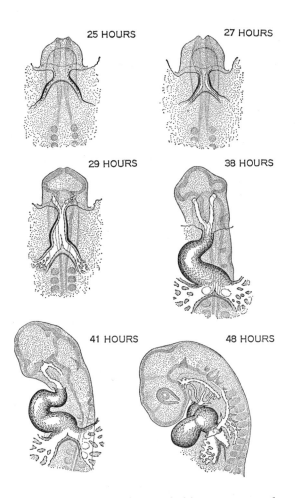

Embryonic chick heart forms from primitive structures that appear about 25 hours after an egg is incubated. Delicate tubes and accompanying ribbons of cells move toward the mid-line and merge into a single tube surrounded by a coat of muscle. First the ventricle forms (29 hours). Then, as the fusion moves rearward, the atrium forms behind the ventricle (41 hours). Finally the "pacemaker" forms behind the atrium. These drawings are adapted from Bradley M. Patten's *Early Embryology of the Chick,* published by the Blakiston Company.

ments usually include cells of all three embryonic layers. If the fragments lack endodermal cells, pulsating masses still develop, but they are not as well organized as those which arise when endoderm is included.

At what point do the cells that are to form the heart assume their chemical identity as heart cells? When do they begin to synthesize the special constituents of heart muscle? One of the first clues came from experiments I conducted at Johns Hopkins University nearly 10 years ago. At that time advances in the field of immunology led Ruth Cooper of Princeton, A. M. Schechtman of the University of California at Los Angeles and other embryologists to utilize antigen-antibody reactions in the study of embryos. The antigen in this case is furnished by that organ in the adult animal which the investigator wishes to trace in the embryo.

An extract of the characteristic proteins of an organ in the chicken, for example, will induce the formation of antibodies when injected into a rabbit. The antibodies that appear in the rabbit blood serum will react in turn with chicken tissues. Since each organ of the chicken contains a mixture of proteins, some of which are found in other tissues of the animal, the rabbit serum will contain a mixture of antibodies. However, by "absorbing" the serum with the various chicken tissues, one can remove all the antibodies except those for the proteins peculiar to the organ under investigation.

By this procedure I prepared a serum that would react only with chicken heart. The "anti-heart" serum gave rise to a clear-cut reaction when it was mixed with an extract of the chick embryo at the primitive-streak stage—several hours before the appearance of the primitive structures of the heart. The reaction showed that the early embryo contains substances identical with or closely related to those of the adult heart. This conclusion was soon supported by the finding that early embryos do not develop a heart when grown on a medium containing anti-heart serum, although the serum does not affect the emergence of other organs. Here was promising evidence that immunological reactions might be used to detect the first appearance of the proteins specific to the formation of heart muscle in the embryo.

We know that three proteins—actin, myosin and tropomyosin—make up about 75 per cent of the total protein of muscle. Myosin extracted from the adult chicken heart contains three molecular components, each of which can elicit antibodies when injected into rabbits. By absorbing the anti-heart-myosin serum with myosin extracted from the leg muscle of chickens we succeeded in preparing antibodies specific for heart myosin, and had thus developed a chemical tool that could distinguish it from the myosin of other types of muscle. We found that this serum did not react with extracts of embryonic tissue before the appearance of the primitive streak. Heart myosin first shows up shortly after prospective mesodermal cells have begun their migration from the periphery of the embryonic disk toward the primitive streak. At the termination of movements through the streak, heart myosin is distributed widely, probably in the mesoderm. (It has not been detected in the endoderm, but its absence from the ectoderm has not been established.)

This distribution surprised us, because embryologists had long believed that a group of cells had to be established at a certain location in the embryo before it could give rise to specialized tissues. Two to three hours after the mesoderm is completed, however, heart myosin is restricted to the heart-forming regions. We found that heart actin, the other component of the contractile fibril, becomes detectable just when the heart-forming regions become demonstrable.

The localization of heart myosin may result from the movement of cells; that is, the cells capable of synthesizing this myosin sort themselves out from among the other mesodermal cells and come together in the heart-forming regions. But another view suggests that the cells outside the two heart-forming regions lose their ability to produce heart myosin through the failure or the inhibition of one or more steps in the synthesis. It is not a simple task to dissociate these phenomena; at such early stages of development cell movements and changes in synthetic processes are closely related.

In searching for a clue to the reason why heart myosin disappears from tissues outside the heart-forming regions, we have tested the synthetic capacities of isolated fragments of the embryo. Embryos were cut in various ways; the fragments were cultured separately and then analyzed for their myosin content with anti-heart-myosin serum. The results were confusing. . . . These experiments have not yielded the hoped-for clarification of the role of location in determining which cells will specialize in the production of heart myosin.

It must be emphasized, however, that we are using antibodies reactive to the heart myosin of adult animals in our effort to detect embryonic heart myosin. The parts of a molecule essential to its physiological function are not necessarily those that combine with an antibody, and the fact that both the embryonic and the adult molecule react with the same antibody does not establish the identity of the two substances. It shows only that the immunologically active groups of the two molecules are identical or have a close similarity. Strictly speaking, our immunological techniques prove merely that certain cross-reacting groups are present. We have not yet determined whether the myosin first detected is adult heart myosin, or a subunit of that molecule, or a complete myosin molecule closely related to but not identical with the adult protein (as in the case of fetal and adult hemoglobin molecules).

It thus appears that in the formation of heart muscle certain of the contractile proteins are synthesized first and later aggregated in the form of fibrils that can contract. As yet we have little evidence on this point outside that furnished by immunological experiments, although some electron microscope studies have also suggested a stepwise organization of the fibril. We cannot, however, be sure of this; in developing skeletal muscle immunological techniques have detected myosin only after the first simple fibrils have formed.

The selective effects of various substances that interfere with normal cell chemistry have produced additional striking evidence that the heart-forming regions are distinguished chemically well before structures of the heart appear. Nelson Spratt of the University of Minnesota has employed certain enzyme inhibitors to show that the metabolic pathways operating in the development of the brain and of the heart differ markedly. In studies I completed recently in collaboration with Lowell Duffey we cultivated early chick embryos in a medium containing traces of the metabolic inhibitor antimycin A, a substance produced by the *Streptomyces* mold. We found that concentrations of this inhibitor as low as .1 microgram per embryo block almost completely the development of the regions destined to form muscle, but leave the developing brain and spinal cord intact. Another metabolic inhibitor, sodium fluoride, has a similar effect. In low concentrations it primarily affects the heart, but at high concentrations it causes the embryo to disintegrate according to a clear-cut pattern starting in the heart-forming regions. At any given stage of development, from the appearance of primitive streak through the establishment of the heart, the locations of the cells destroyed by sodium fluoride coincide with the sites that have the greatest capacity to form heart muscle, and with the areas that have the greatest capacity for the synthesis of actin and myosin. Thus the primary forces in the formation of the heart seem to be operating almost at the very outset of embryonic development.

With the present rapid advance of biochemical techniques we should soon be able to state accurately when and where a given protein is first formed. To ensure continued progress, however, the biochemist must learn from embryology about cellular organization, the role of the cell surface and the interactions of the cells and tissues. One of the most effective tools of the experimental embryologist has been microsurgery. Now "chemosurgery" provides another means for deliberate intervention in the pattern of development and for the alteration of its course under experimental control.

III
PROBLEM DEFINITION AND PROBLEM SOLVING

A brilliant biologist whose research has dealt largely with mutations and inheritance in bacterial cells, Dr. Evelyn Witkin
wrote "Mutations and Evolution" for a special science edition of
Atlantic Monthly. The article appeared in the issue of October
1957. In it she recounts with grace and clarity how the science of
genetics, a branch of study not yet born in Darwin's day, has carried
forward our understanding of evolution.

Dr. Witkin was born in New York City in 1921. She was
graduated from New York University in 1941, and later received
her master's and doctor's degrees at Columbia University.

She began her association with Carnegie Institution while
still a graduate student at Columbia, choosing to do research for her
doctoral thesis at the Institution's Department of Genetics at Cold
Spring Harbor, Long Island, New York. While working toward
her degree she was appointed a research assistant at the Department, and participated in a research project there for the Office of
Scientific Research and Development.

In 1946 she received a grant from the American Cancer
Society which enabled her to extend her research collaboration with
Carnegie Institution geneticists, and in November 1949 she was
appointed a junior staff member of the Institution. For the next five
years she continued her research in bacterial genetics at Cold Spring
Harbor.

Two of her most notable achievements during that period
were the discovery of spontaneous mutations causing bacteria to
become resistant to radiation damage, and the demonstration that
the "nucleus" that is visible in the bacterial cell actually carries its
hereditary material.

In 1955 Dr. Witkin transferred her research program from
Carnegie Institution's Department of Genetics to the Downstate
Medical Center of the State University of New York, where her
husband is Professor in the Department of Psychiatry, and where
she is now Associate Professor in the Department of Medicine.
She has, however, since continued her relationship with the
Carnegie Institution as a research associate.

140

Evelyn M. Witkin

MUTATIONS AND EVOLUTION

From *The Atlantic Monthly,* October 1957.
Copyright, 1957, by The Atlantic Monthly Company, Boston, Mass. 02116.
Reprinted with permission.

When Charles Darwin began his voyage around the world aboard H.M.S. *Beagle,* he shared with his contemporaries the almost unquestioned belief that every species of plant and animal then inhabiting the earth had originated in a separate act of creation. No other way had ever been found to explain the exquisite adaptations of structure and behavior by which each form of life seems so perfectly designed for its place in nature. By the end of the five-year journey, an altogether new and startling idea had begun to develop in the mind of the young naturalist. Today, less than a century after the publication of *The Origin of Species,* the theory of evolution has long been accepted as a fact of life.

The brilliance of Darwin's insight lay in his integration of two simple and seemingly unrelated biological truths, and in his projection of their inevitable consequences on a vast scale of time. One was that the individual members of a species are not all precisely alike, the differences among them tending to be inherited. The other, somewhat less obvious, was that the infinite expansion of populations is checked by limitations in the availability of food, and by other restrictive conditions of life. It follows directly, reasoned Darwin, that any inheritable trait that enhances the survival and fertility of an individual will be "naturally selected"—that is, will be transmitted to a larger fraction of the population in each succeeding

141

generation. In this way, by the gradual accumulation of adaptive variations, the species now existing have evolved from earlier and more primitive progenitors, and owe their intricate mechanisms of adjustment not to purposeful planning but to the impassive operation of natural laws.

In the great upheaval of scientific thought that followed the announcement of the theory of evolution, the phenomena of heredity and variation were suddenly thrust into the forefront of biology. Almost nothing was known of the way in which hereditary differences arise, and of the mechanisms of their transmission, but Darwin foresaw the development of a "grand and almost untrodden field of inquiry" in which the causes of variation and the laws of heredity would be discovered. Even as Darwin called upon the future to solve the mysteries of inheritance, Gregor Mendel was laying the foundation for the new science of genetics. Genetics has contributed richly to the synthesis of facts and ideas from almost every branch of the natural sciences that has been built upon Darwinism. As the diverse and intricate mechanisms of evolution have come to be understood, it has grown increasingly certain that the raw materials upon which they depend are the mutations of genes.

The hereditary endowment of a plant or animal is now known to be determined by a very special kind of material found primarily in the threadlike chromosomes that may be seen under the microscope in the nucleus of the cell. The invisible elements of which this material is composed, the genes, were once regarded as discrete particles strung along the chromosome like beads. Recent evidence has modified this concept considerably, and many geneticists now think of genes as chemically differentiated regions of the chromosome, not necessarily separated one from another by definite boundaries, but each having a distinctive structural pattern from which it derives a highly specific role in the metabolism of the cell.

Every cell in the body contains a set of chromosomes

and genes, descended directly by a long line of cell divisions from the set originally constituted in the egg cell at fertilization. The human embryo develops into a person, rather than into a tree or an elephant or a monstrosity, because the material carried in its chromosomes, its constellation of genes, initiates and guides a marvelously coordinated sequence of reactions that leads inevitably, under normal conditions, to the differentiation and growth of a human being.

Throughout the life of the individual, the genes continue to exert their control over the complex chemistry of the cells and tissues of the body. As older tissue is gradually replaced by new tissue in the mature person, the food that is consumed is converted quite specifically into more of the very same individual, even though an identical diet, fed to a dog, would be transformed into more dog. We are a long way from understanding just how genes direct the manifold activities of living systems, but we know with growing certainty that the range of possible responses of any cell or organism to the conditions it may encounter is largely gene-determined.

All the members of our species have in common the basic genetic make-up that sets us apart from other forms of life. Nevertheless, no two individuals, with the exception of identical twins, have exactly the same heredity, which is another way of saying that every person possesses a unique pattern of chromosomal genes. Differences in skin pigmentation, eye and hair color, stature, and facial features are familiar hereditary traits by which individuals and groups of individuals differ from one another. These and the host of other inherited variations, from fingerprint patterns to blood types, are manifestations of the differences that exist in the structure and arrangement of the genic material.

Some hereditary variations, such as eye color, are known to depend upon differences in the state of a single gene. This does not imply that one gene, all by itself, is responsible for the formation of blue or brown pigment in the iris of the eye. It means that a change in this particular gene can alter the integrated functioning of the whole gene system so as to result in the production of a different kind of pigment. Other characteristics, such as height, depend upon the states of a relatively large number of genes.

Genes do not exist in a vacuum. They are always present in an environment that must be taken into account in understanding how they work. The environment within the cell and within the organism, and the more unpredictable environment outside, are intimately bound up with the functioning of genes and have varying degrees of influence upon the ultimate expression of heredity. A trait or characteristic is not, in itself, inherited. That which is determined by genes is the capacity to produce certain traits under certain conditions.

In the case of eye color, this distinction may seem unimportant, since an individual having the genetic constitution for blue eyes will have blue eyes under any environmental conditions. Its meaning becomes evident, however, when we consider inherited characteristics that are more directly responsive to environmental variables. The Himalayan rabbit is a case in point. This rabbit has a pattern of white fur, with black fur at the extremities (ears, tips of paws, tail), and this pattern is passed along from generation to generation. If a patch of white fur from the back of such a rabbit is shaved off, and the new fur allowed to grow back while the animal is kept in a cool place, it will grow in black instead of white. Thus it is not the pattern itself that is inherited, but the capacity to produce black pigment at low temperatures and not at higher temperatures. Since the temperature at the extremities is normally lower than that of the rest of the body, the typical Himalayan pattern is obtained. Similarly, although stature is basically under the control of genes, it can be influenced significantly by nutritional factors.

Genes are remarkable not only for the way they direct the intricate pathways of metabolism and development. They have, in addition, unique properties that give them special importance in biology, as the raw materials not only of evolution but probably of life itself. Genes have the ability to organize material from their surroundings into precise copies of their own molecular configurations, and they exercise this power every time a cell divides.

They are also capable of undergoing structural changes, of mutations; and once such a change has occurred, it is incorporated into the copies that the gene makes of itself. A single unit having these properties, and having also the ability to aggregate with other such units, would possess the essential features of a living being, capable of unlimited evolution through the natural selection of variant forms and combinations most efficient in reproducing themselves. Many biologists believe that life may have originated with the accidental formation of "naked genes," organic molecules able to duplicate their own structure, and their variations in structure, from materials available in the environment. . . .

One of the most important recent advances in genetics is the evidence that the definitive properties of the genes can be accounted for by the theoretical structure and behavior of the molecules of compounds known as deoxyribonucleic acids, or DNA. Chromosomes contain large amounts of DNA. Its molecules are very big, as molecules go, built up in long chains from only four kinds of simple chemical building blocks. The order in which these units occur, and the number of repetitions of similar groupings, are thought to be the basis of the specific activity of different regions of the chromosome—in other words, of genes. The study of the properties of these molecules provides a way to explain the mechanism by which genes duplicate themselves and reproduce the variations that they may undergo.

Mutations, as has already been suggested, are considered to be changes, on the molecular level, in the structure or organization of genes. A mutation in any gene is likely to be reflected in a modification of its contribution to the delicately interwoven pattern of control exercised by the whole constellation of genes, and may be detected by its effect on some physical or metabolic characteristic of the organism.

Mutations, in nature, are rather rare events, occurring usually with frequencies of from one in a thousand to one in a billion gene duplications. They have an extremely wide range of effects, from fatal disturbances of normal development to barely perceptible reductions of life expectancy, from striking changes in appearance to slight alterations of metabolism that can be detected only with sensitive laboratory instruments.

Mutations in man are responsible for the kinds of hereditary differences we have already discussed, and can produce, as well, such effects as early fetal death, stillbirth, diseases such as hemophilia and sickle cell anemia, color blindness and harelip. It seems quite possible that cancer, leukemia, and other malignant diseases may originate by the occurrence of mutations in body cells other than the reproductive cells.

Although the overall frequency of mutations can be increased considerably by exposure to radiations and a variety of chemicals, there is ordinarily no relation between environmental conditions and the *kinds* of mutations that occur. Mutations of all sorts arise in natural populations, with low but regular frequencies, in a way that is best explained by considering them to be the consequences of accidental molecular rearrangements, occurring more or less at random in the genetic material. X rays and other kinds of high-energy radiations increase the probability that these accidents or mutations will occur, but we do not know with certainty the causes of so-called "spontaneous" mutations. Natural radiations, such as cosmic rays, undoubtedly cause a fraction of them, but it has been estimated that the intensity of natural radiations is not sufficient to account for all the mutations that occur in plant and animal populations.

Darwin believed that the inheritable variations upon

which natural selection acts are caused directly by the influence of the conditions of life upon the organism, or by the effects of use and disuse of particular body parts. Although he appreciated the difficulty of explaining how the environment can provoke appropriately adaptive modifications, and how such changes can be incorporated into the reproductive cells so as to be inherited, it seemed at that time even more difficult to imagine that they could arise by chance. How, then, does modern genetics propose that the orderliness of evolution can follow from accidental variations in the molecular structure of genes, occurring without relation to the demands of the environment?

We do not need to rely upon speculation to answer this question. The study of evolution has moved into the laboratory, and while it is not possible to duplicate here the kinds of changes that have required millions of years in nature, the elementary steps of evolution can be analyzed. For this purpose, the use of bacteria presents many advantages. This is particularly true since the mechanisms of heredity and variation, wherever studied in the plant and animal kingdoms, seem to be fundamentally alike. Genes and mutations are much the same, in their basic behavior, whether they are investigated in fruit flies, in maize plants, in man, or in microorganisms.

The bacterium *Escherichia coli,* a rod-shaped, one-celled organism normally found in the human intestinal tract, is widely used in research on heredity. It divides every twenty minutes under optimal conditions, and a single cell, placed in one cubic centimeter of culture medium, will produce overnight as many descendants as the human population of the earth. The recent discovery of a sexual process in this organism, as well as in some other kinds of bacteria, has made it possible to interbreed different strains and to apply many of the classical methods of genetic analysis that were developed in the study of higher forms. *Escherichia coli* is an ideal vehicle for the experimental study of "microevolution."

In the laboratory, a strain of this bacterium can be

maintained almost indefinitely, under constant conditions, without undergoing any appreciable change in its characteristics. When the environment under which the bacteria are grown is changed, however, in a way that is somehow detrimental to the population, it will often adapt itself rapidly and effectively to the new conditions.

A good example of the way in which a bacterial culture may adapt to an unfavorable environment is the reaction of *Escherichia coli* to streptomycin. Most strains of this bacterium are sensitive to streptomycin, and are unable to multiply in the presence of even very small amounts of the antibiotic. Sensitivity to streptomycin is an inherited trait and is transmitted, unchanged, through countless generations. If a high concentration of streptomycin is added to the culture tube in which a sensitive strain is growing, the outcome depends upon the size of the population at the time. If the number of bacteria in the tube when the antibiotic is added is relatively small (a hundred or a thousand), multiplication will stop at once, and no further growth will take place in the tube, no matter how long it is incubated. If the population is large (a hundred million bacteria or more), the addition of streptomycin will arrest multiplication sharply, but incubation of the tube for a few days will almost always result in the ultimate appearance of a fully grown culture containing tens of billions of bacteria. When the bacteria in this culture are tested, they prove to be completely resistant to streptomycin, and are able to multiply vigorously in its presence. Further, we find that resistance to streptomycin is a stable, hereditary characteristic, transmitted indefinitely to the descendants of these bacteria.

Thus, by exposing a large population of streptomycin-sensitive bacteria to a high concentration of the antibiotic, the emergence of a genetically resistant strain can be brought about. This, indeed, is a strikingly adaptive change, and at first sight it may seem to substantiate the old idea that the environment can cause useful modifications that are then inherited. The careful study of the events leading to the appearance of a streptomycin-resistant strain proves without doubt that this is not so.

It can be readily demonstrated, first of all, that the adaptation to streptomycin does not come about by the mass conversion of the entire sensitive population, but rather is the result of the selective overgrowth of the culture by a few individuals that are able to multiply in its presence, while the division of the rest of the population is inhibited. It is for this reason that adaptation occurs only when the exposed population is large enough to contain at least one such individual. The critical question is this: how did these rare individuals acquire the properties that enabled them and their descendants to multiply in the presence of streptomycin?

This question has deep roots in biological controversy. It recalls, in a new form, the arguments over Lamarck's idea that modifications of the individual caused by environment can be inherited by descendants. Although Lamarckism has long since been disproved to the satisfaction of most biologists by repeated demonstrations that such inheritance just doesn't happen, the idea has persisted in bacteriology until very recently that microorganisms are somehow quite different from other plants and animals, and that permanent hereditary changes of an adaptive kind can be produced in bacteria directly as a result of the action of the conditions of life.

Two alternative hypotheses can be considered in planning experiments to determine the true origin of streptomycin-resistant variants. The first is that a small number of initially sensitive bacteria were modified as a direct result of the action of streptomycin, thereby acquiring permanent resistance. This would be an example of an adaptive hereditary change caused by the environment, as Darwin envisaged the origin of most hereditary variations. The second possibility is that the resistant individuals had already acquired the properties necessary for resistance *before* coming into contact with streptomycin, as a result of a mutation during the normal division of the sensitive population. In this case, the role of the antibiotic would be entirely passive, providing conditions that favor selectively the multiplication of those rare individuals present in the population that are already equipped, by virtue of the previous occurrence of a chance rearrangement of a particular gene, to withstand its inhibitory action.

During the past fifteen years, a great many experiments have been designed and conducted in a number of laboratories for the purpose of determining which of these hypotheses is correct. They have established beyond doubt that the second one is right, and that streptomycin-resistant variants originate by mutation, at a very low rate, during the growth of sensitive strains that have never been exposed to streptomycin. The proof depends upon the demonstration that the very first generation of resistant individuals in a culture to which streptomycin has just been added already consists of related family groups, or clones, in just the way that would be predicted if their resistance were the consequence of a hereditary change that had taken place some generations back.

The development of resistance to streptomycin illustrates the way in which mutations provide the basis for adaptive changes in bacterial populations. Actually, any culture of *Escherichia coli*, apparently quite homogeneous when hundreds or even thousands of bacteria are compared, contains within it rare variants that differ from the predominant type in one or more of countless ways. When a suitable selective environment is provided, it can be shown that a culture contains mutants resistant to many antibiotics, to the action of radiation, to all sorts of chemicals that inhibit particular steps in metabolism—mutants that differ from the standard type in the sugars they can ferment, in their rate of growth, in the complexity of their nutritional requirements, in their antigenic properties, and in almost any characteristic for which a method of detection can be found.

In every case that has been carefully studied, these differences are found to originate without any contact with the conditions under which they happen to be advantageous, and their rates of occurrence are ordinarily not increased by such contact. This is true not only in bacterial cultures, where mutations can be demonstrated rapidly and dramatically. Natural populations of other plants and animals, including man, are known to contain mutations of many kinds that occur with no apparent causal relation to the conditions of growth.

Thus, in a way that Darwin could not have surmised, chance, through mutation, plays a most important part in evolution. It would be difficult indeed to imagine how a species could long survive, or progress in evolution, if it were dependent for its flexibility upon variations directly caused by the conditions of life. Quite aside from the fact that modifications produced in this way are not inherited, except in very special cases, it would require the intervention of some purposive and prescient agent to guarantee that previously unencountered conditions could typically provoke in the organism just those responses that are required to enhance adjustment.

Of course, the occurrence of a diversity of mutations in populations of bacteria and other organisms does not necessarily equip them to meet successfully every environmental challenge. Some strains of bacteria, for instance, are unable to adapt to streptomycin, since their spectrum of mutations does not include the particular modification of metabolism that is required for streptomycin resistance. Furthermore, since there are limits to the range of conditions that can support life, any sufficiently drastic changes, such as those that would take place in the center of a hydrogen bomb explosion, are not likely to prove conducive to the survival of any living thing.

Even within the range of more tolerable conditions, the suddenness of change is sometimes more decisive than its magnitude. For example, the bacterium *Escherichia coli* can be made resistant to streptomycin, penicillin, and chloromycetin, if the mutants resistant to each of these antibiotics are selected sequentially, but such a triply resistant strain cannot be obtained if the sensitive strain is exposed simultaneously to all three agents. This is explained by the negligible probability that any one individual in a finite population will have undergone mutation in three particular genes, each of which mutates very infrequently and independently of the others.

Observations of this kind, incidentally, although originally made in laboratories of genetics, have found important applications in medical practice. Many people who have used antibiotics to combat infection have had the experience of dramatic relief of symptoms, only to be followed within a few days by a recurrence, this time failing to respond to the same antibiotic. Sometimes this can be explained by selection of a variant, present in the infecting population of bacteria, that is resistant to the antibiotic and that has its chance to multiply once the sensitive population is eliminated by the first round of treatment. In some cases, a physician will recommend the use of a combination of two or more unrelated antibiotics simultaneously, knowing that mutants resistant to more than one such drug are much less likely to be present. While the use of combinations of antibiotics is not always feasible for medical reasons, under certain conditions it has effectively prevented the occurrence of relapses caused by selection of resistant variants.

There is, of course, much more involved in the complicated saga of evolution than the simple picture of mutation and selection that accounts for bacterial adaptation to streptomycin. Nevertheless, the continuity of life from its first stirrings, and its steady progress toward higher levels of organization, has depended, and continues to depend, upon the reservoir of adaptive responsiveness that is provided initially by the mutations of genes.

Why, it may be asked, if mutations are the source of evolutionary progress, do we hear so much about the genetic dangers of radioactive fall-out, overexposure of the reproductive organs to clinical radiations, and the heightened radiation levels of the atomic age? We know that radiations increase considerably the frequency with which mutations of all sorts occur. Mutations, in themselves, are neither good nor bad. Streptomycin resistance is good for *Escherichia coli* in the presence of streptomycin, but when the antibiotic is removed, many of the resistant mutants are unable to grow, some of them actually requiring streptomycin for growth. Similarly, radiation-resistant mutants are at a distinct advantage in the presence of ultraviolet light or X rays, yet, in competition with the sensitive form when no radiation is present, they die out rapidly. At any stage in the history of a species, under natural conditions, the mutations that are occurring have undoubtedly occurred before, and most of those that are advantageous under the conditions then prevailing have already been established as part of the predominant gene complex. Thus most mutations are bound to be harmful in some way; the most frequently occurring mutations in the fruit fly are known to be those having lethal effects. Increased mutation rates as a result of exposure to unnatural amounts of radiation, therefore, are likely to be injurious, not only to the individual progeny of particular people, but to the vigor of mankind.

While the genetic hazards of radiation are of most immediate concern, there are more positive implications of the new knowledge of genetics and evolution for the future of humanity. The degree of control that has been achieved over environmental forces, and over the constitutional infirmities that would otherwise reduce the chances of survival and procreation of a significant segment of mankind, has already weakened the hitherto unchallenged power of natural selection. If man should one day choose to put to use the far greater power of his conscious and purposeful intervention, his biological future will be shaped by his own hands. There are still undreamed-of possibilities in the multipotent clay that is his to mold.

D*r. C. Stacy French, Director of Carnegie Institution's Depart-*
ment of Plant Biology and one of the leaders in research on
photosynthesis, wrote this article on "Photosynthesis" as a chapter
in the book This Is Life, *published in 1962 by Holt, Rinehart and*
Winston. For reasons of space it is here presented in somewhat
condensed form.

As Director of the Department of Plant Biology at Stan-
ford, California, Dr. French presides over an organization devoted
to fundamental research on the physiology, genetics, and function-
ing of plants. The Department operates generally in two areas:
research on photosynthesis, especially the function and significance
of pigments in this process, and the field called experimental
taxonomy, the goal of which is the understanding, not of a single
plant process, but of the whole chain of mechanisms that determine
plant evolution.

Dr. French joined the Carnegie Institution as Director of
the Department of Plant Biology in 1947, succeeding Dr. Herman
A. Spoehr. He was born in 1907 at Lowell, Massachusetts, and
attended Harvard University, specializing in physiology, where he
received the degree of B.S. in 1930 and the Ph.D. in 1934. The
following year he spent as a research fellow at the California In-
stitute of Technology. In 1935–1936 he was a guest investigator
at the Kaiser Wilhelm Institute in Berlin.

The following year he returned to Harvard as Austin Teach-
ing Fellow in the Department of Biochemistry of the Harvard
Medical School. From 1938 to 1941 he was research instructor in
the Department of Chemistry of the University of Chicago. From
1941 to 1946 he was assistant and associate professor of botany
at the University of Minnesota, leaving that position to come to
the Carnegie Institution.

His research interests have centered mainly on photosyn-
thesis, particularly the nature and function of the pigments that
participate in this vital process. He has written numerous technical
papers in this and related fields. Dr. French has a special talent not
only for research, but for the devising of ingenious apparatus to
further his investigations. Among these have figured conspicuously
a new type of spectrophotometer for recording the first derivative
of absorption spectra, an automatic machine for plotting the rate
of photosynthesis in algae at different wavelengths of light, and a
graphical computer for the analysis of data.

154

C. Stacy French

PHOTOSYNTHESIS

From *This Is Life,* published by Holt, Rinehart and Winston, Inc., 1962.

The food we eat and the oxygen we breathe are both formed by plants through the process of photosynthesis. The power to drive the photosynthetic reaction comes from sunlight absorbed by chlorophyll in plants. Although all life depends directly or indirectly on photosynthesis the chemical nature of this remarkable energy-converting process is not at all well understood. No known chemical system can be made to serve as substitute for this ability of plants to turn carbon dioxide into organic matter and free oxygen, using energy from sunlight. Furthermore, chlorophyll, in its functional form in plants, is a different substance from extracted chlorophyll. The chemistry of chlorophyll, in its natural state of combination as a protein complex, is almost completely unknown, even though that material is the most obvious organic substance on earth. The reason for our ignorance about this extraordinary natural process is not so much due to lack of scientific effort as to the inherent complexity of the system. . . .

One of the early experiments in plant physiology was a very simple and beautiful experiment with a willow tree, undertaken by van Helmont (1577–1644). Somewhat ahead of his time, van Helmont had the idea that one could find out about things better by making measurements and observations than by studying the literature. Scientific literature then consisted almost entirely of Aristotle's writings, which even at that time were over a thousand years old.

Van Helmont's historical experiment was to weigh a small willow shoot and also a tub of soil. He then planted the willow in the soil and watered it carefully for five years. When he removed the tree it had gained 164 pounds while the soil had lost only a few ounces. Therefore he came to the conclusion that the extra weight of the tree had come from the water. This conclusion, like most scientific conclusions of our own day, was partly right and partly wrong. We know now that, in addition to the water, much of the weight of this tree had come from atmospheric carbon dioxide. In van Helmont's time no one knew that such a thing as carbon dioxide existed. His experiment was the beginning of the story of photosynthesis — the story of how plants produce their own food and ours by using carbon dioxide and sunlight.

In those days the rate of scientific progress was somewhat less bewildering than it is today. In fact, 125 years went by before the next experiment bearing on our subject was tried. In 1774, Joseph Priestley published a description of some experiments that uncovered a new and startling effect. He found that air in which mice had been kept until they died was in some strange way regenerated and again made usable for the mice by the insertion of green plants. It is obvious to us today that Priestley's mice had used up the oxygen and produced carbon dioxide, and that the plants converted the carbon dioxide back to oxygen again through the process of photosynthesis. In Priestley's time oxygen and carbon dioxide were unknown, so that his experimental observation lacked a rational scientific explanation. Nevertheless, it opened the way to a very important development in understanding the carbon cycle in nature.

In work with plants during Priestley's time a major emphasis was placed on their description and classification. It happened to follow the most productive period of Linnaeus, who proposed the first workable system of classification. Although Linnaeus himself performed and strongly advocated experiments with plants, many of his successors paid little attention to this aspect of his work. They had become too engrossed in using the new tool provided by the simplification in classification.

This enthusiasm for classification, the academic counterpart of stamp collecting, has remained a major part of botanical research since the time of Linnaeus. Even to this day the study of how plants function, as compared with the study of their shapes and taxonomic relationships, is generally a small part of elementary botany courses. The process of photosynthesis, however, has continued to attract the attention of scientists with many different kinds of training and experience.

A Dutch physician, Ingenhousz, who traveled widely about Europe and enjoyed a prosperous career as a Royal physician, somehow found time in his busy and urbane life to carry on experiments on photosynthesis. The great contribution Ingenhousz made in about 1780 was to find that light was necessary for the purification of air by plants. Three years later, Senebier made another great advance in the understanding of the basic principles of photosynthesis in terms of plant physiology. He found that only the green parts of plants were active and, furthermore, that the volume of good air plants produced was dependent on the amount of bad air they received. In the next year, 1784, Lavoisier, who discovered oxygen (and later lost his head over politics), worked out the composition of carbon dioxide and the nature of its formation by burning carbon compounds. The basic chemistry was now available for a clear understanding of the process of photosynthesis, which is the conversion of carbon dioxide and water into organic material.

Although the first textbook of plant physiology was published in 1795 by the Danish botanist C. G. Rafn, the entire process of photosynthesis was not clarified and described in chemical terms until 1804. In that year de Saussure showed by gas analysis that carbon dioxide was used up during the formation of organic matter and oxygen was produced.

In 1818, Pelletier and Caventou gave the name of "chlorophyll" to the green stuff that could be extracted from plants by alcohol. Yet even at the present time no one knows how chlorophyll works in photosynthesis. In 1840, the German agricultural chemist Liebig clearly showed that it was the inorganic salts in soil that were necessary for plant growth and not the organic matter. He understood that humus in the soil had its source in plants, in contrast to the old idea that plants were dependent upon the humus in the soil for their growth.

The quantitative aspect of the utilization of light by plants was first considered seriously in 1845, when Robert Mayer formulated the law of conservation of energy. Although this law seems clearly self-evident now, it was a most important discovery in its time. The concept that energy can be changed from one form to another but does not disappear had an immediate bearing on the subject of photosynthesis in plants. Mayer realized that photosynthesis is a conversion of light energy to the form of stabilized chemical energy stored in organic substances.

Thus after 75 years, from the time when Priestley discovered that air used up by mice could be freshened by leaving plants in it the energy conversion concepts of Mayer completed the clarification of the overall picture of just what photosynthesis is and what it does. Photosynthesis can be summarized by the equation

$$6CO_2 + 6H_2O + 672,000 \text{ cal} \rightarrow C_6H_{12}O_6 + 6O_2$$

This says that 672,000 calories of absorbed light energy convert 6 moles of CO_2 and water into 1 mole of carbohydrate and 6 moles of oxygen. The equation only tells what goes in and what comes out. This basic chemical reaction of photosynthesis was as well recognized a hundred years ago as it is today. The problem then became the more difficult one of finding out what different chemical steps take place to produce this overall result. Today we ask not what photosynthesis is, but rather how it is carried on. We are still unable to duplicate photo-synthesis by any known chemical system.

Up until a hundred years ago, botanists were chiefly concerned with classification of plants, and so knowledge of photosynthesis was advanced more by chemists than by botanists. After about 1860, however, plant physiology as a part of botany began to develop under the influence of Sachs, Pfeffer, Timiryazev, and others who were primarily students of plant life rather than of chemistry.

One of the important discoveries regarding the photosynthetic system of plants was made by a physicist. This was Stokes, an Englishman who was concerned with the process of fluorescence — that is, the re-emission of absorbed light as a particular color determined by the chemical structure of the material. Chlorophyll is highly fluorescent, and the effect is easy to see. A leaf mashed in alcohol gives a green solution that can be clarified by filtering through cloth. This green extract will show red fluorescence if placed under a bright light in a moderately dark room (a slide projector is a good light source). Stokes, in purifying chlorophyll to study its fluorescence, found there were two forms of chlorophyll, which he called chlorophyll *a* and chlorophyll *b*. About one fourth of the chlorophyll in plants is chlorophyll *b*. This discovery was made in 1864, and we have only recently begun to have a reasonably good idea as to why plants need to have both kinds of chlorophyll.

From 1895 to 1911 a great deal of research on the rate of photosynthesis was carried on at the University of Cambridge by Blackman and his group. They studied the effects of environmental factors, such as light intensity and temperature, as well as internal factors controlled by the nature and previous treatment of the plant. The effects of these different factors were found to be interrelated, and the level of one variable was seen to determine how the other variables influence the speed with which plants carry on photosynthesis.

Following the Blackman period Willstätter and Stoll produced two monumental volumes, one on chlorophyll in

1913 and one on the process of photosynthesis in 1918. Both books were based on many years of intensive laboratory investigation. About 1920, Warburg introduced improved measuring methods and the use of unicellular algae, thus vastly extending the quantitative study of the process. Most present-day investigations of photosynthesis and of pigment function reflect the methodology, the standards of precision, and the clear thought brought to bear on photosynthesis by Warburg.

Although Priestley's last days were spent in the United States, photosynthesis otherwise remained an entirely European research activity until 1911, when Spoehr, who with Warburg had studied under Emil Fisher in Berlin, started his career in this country. In 1926 Spoehr published a monograph summarizing everything known about photosynthesis up to that time. His was to remain the major source of reference on the subject for twenty years. The next American, Emerson, returned from Warburg's Berlin laboratory in 1927, and for some years photosynthesis research in this country was almost entirely the province of Spoehr, Emerson, and their collaborators.

About the middle of the 1930 to 1940 decade the European tradition of photosynthesis investigation was reinforced and diversified by the immigration of such men as van Niel, Franck, Gaffron, and Rabinowitch, and later Kok. They, as well as Spoehr, Emerson, and Burk, established laboratories, trained students, and developed active groups of investigators in the United States. European interest also continued to expand. . . .

Research on photosynthesis has kept many men busy for a long time, and workers from very different scientific disciplines have been concerned with this basic process. It is worthwhile to realize that the recent new discoveries reported in newspapers usually represent only a small amount of information added to a large pre-existing accumulation of knowledge; it is this knowledge that endows the new discoveries with significance.

The Role of Photosynthesis. The energy by which all animals live is generated by the oxidation of plant-produced foods—either directly by eating plants, or indirectly by eating plant-fed animals. This oxidation of organic compounds by respiration or fermentation gives off carbon dioxide to the air. Man's use of organic matter produced by plants ages ago and stored as coal or oil also continuously pours CO_2 into the air. Yet despite the enormous CO_2 production on a worldwide scale, the concentration of CO_2 in air has remained very nearly constant since it was first measured. Evidently the total rate of CO_2-consuming photosynthesis is just about in balance with the total CO_2 production over the whole earth. How does it happen that the two opposing processes, going on entirely independently of each other, remain so nearly equal in their overall volume of production?

A partial answer to this question illustrates a basic principle of biology: the automatic control of environmental factors within or around living things at a level favorable to their survival. Fortunately for the continuance of life on earth, the rate of photosynthesis of all plants is nearly proportional to the CO_2 concentration prevailing in air. Because of this relation, the level of CO_2 in the atmosphere is very hard to change appreciably. In other words, the more CO_2 that is available, the more rapidly it is turned back into oxygen. Conversely, the lower the CO_2 concentration, the smaller its conversion rate to oxygen.

The regulation of the atmospheric CO_2 level by photosynthesis is not at all thoroughly understood. It is, nevertheless, of great importance in affecting the earth's temperature and hence, by variation of the polar ice caps, in determining the level of the oceans.

Whether the current explosion of the human population will have a serious effect on the CO_2 level of the air, or whether worldwide photosynthesis will increase enough to offset the imbalance, has not been given much consideration. However, the total of the earth's food production by photosynthesis is a matter of serious concern to anyone attempting to compile a balance sheet showing food-supply potential and human population estimates in the near future. The fact that well over half the people of the world are hungry all the time, even now, is often brushed aside by writers who have particular motives for obscuring basic facts about the overpopulation problem. Because all food is derived from photosynthesis either directly or indirectly, an understanding of this process is of basic importance to humanity. Yet direct application of research in photosynthesis to increased food production is still a vague hope rather than a present reality. . . .

The Efficiency of Photosynthesis. One of the more obvious questions one might ask about photosynthesis is: How efficient a process is it? This question can be put in still another way: How efficiently does a plant growing out-of-doors use the energy available from sunlight?

The wavelength of much of the energy is too long to be absorbed by chlorophyll, and this energy is wasted. About 20 percent of the incident energy is reradiated as long-wavelength infrared, and about 30 percent of the light goes right through the leaf, unabsorbed, and without having any effect at all. About half the energy of sunlight falling on a normal leaf is used up as heat in the evaporation of water. The answer, then, is that only about 2 percent of the total sunlight is used for photosynthesis by crop plants growing in direct sunlight.

One reason for this low efficiency is that sunlight is very bright and leaves cannot take advantage of very bright light. Furthermore, the small amount of light that is actually used for photosynthesis goes to make a whole plant, not only the useful parts: a large fraction of a typical plant consists of inedible fibers. Finally, much of the energy captured by photosynthesis is used up again in respiration.

If a crop is eaten as a vegetable the food is converted directly into animal or human tissue, although with a low efficiency. About 90 percent of the energy in the crop fed to cattle is wasted in producing the animals. This means that eating meat gives us only about 10 percent of the energy which was originally in the feed used in raising the meat. Underdeveloped and overpopulated countries are therefore primarily bound to vegetarianism.

A second measure of photosynthesis efficiency is the maximum yield of photosynthesis obtainable from plants under their optimum conditions. By shining beams of light of measured intensity into suspensions of single-celled algae like *Chorella,* the maximum obtainable efficiency of photosynthesis has frequently been determined. In measurements of this type it is convenient to express the light intensity, not in terms of energy, but in number of quanta — the smallest unit package of light. This is done so that the results may be compared directly with the number of carbon dioxide molecules reduced. This experiment was carried out by Warburg and Negelein in Berlin in 1922. They found that it took about 4 quanta of light to reduce a carbon dioxide molecule. This experiment was repeated in about 1938 by Emerson and Lewis, who found a requirement of 8 to 12 quanta per CO_2. In spite of a great number of investigations by many different people, the results of such measurements still vary widely for reasons that are not yet clear.

To simplify the subject of photosynthesis it is convenient to imagine the part of a leaf that does the work as a black box. Our basic question is: What does the black box contain and how do its components work? Our investigation is hindered by the fact that if, figuratively speaking, we pry off the cover to look inside our box its wheels stop turning. We can, however, pour carbon dioxide and water into this box and, while shining light on it, take out carbohydrates and oxygen. We can study the way things go on in the black box by making precise measurements of the rate of CO_2 uptake, O_2 evolution, and product formation. Thus by using water labeled with the O^{18} isotope it has been found that the oxygen of photosynthesis is derived from the water, not from the CO_2. The study of the way in which these measurable photosynthetic rates depend on the temperature of the box, the light intensity supplied, and the concentration of CO_2 fed to it has been a major part of photosynthetic research. Many theories regarding the mechanisms within the box have been developed from observations of the way these controllable factors influence the rate of photosynthesis.

The detailed study during the last fifty years of the rate of photosynthesis under different conditions has been of no help in identifying the type of compounds involved, but it has furnished some information as to the complexity of the process. For instance, if we measure the amount of photosynthesis produced by a leaf or a suspension of algae at different light intensities, it is found that the photosynthetic rate increases in proportion to intensity while the light is weak. However, when the light is strong, the addition of more light does not produce any more rapid photosynthesis. This means that there must be some step, not dependent on light, whose rate becomes limiting when the strictly photochemical part of the mechanism is going as rapidly as it can. The effect of varying the color of the light depends on the pigments that are present and so will be discussed after the pigments have been described.

CO_2 concentration influences the rate of photosynthesis in much the same way as does the light intensity. At low pressures of carbon dioxide, such as are present in normal air (about 0.03 percent), the rate of photosynthesis increases proportionately with the supply of carbon dioxide. Further increases above 0.2 percent CO_2 in the air have no influence on photosynthetic rates until toxic effects appear, at concentrations of about 10 percent.

When the light intensity is high enough so that the photochemical part of the process is not limiting the rate, the effect of temperature on photosynthesis is much like its effect on other enzymatic reactions. Thus in bright light the photosynthetic rate increases exponentially with temperature in the low temperature range (Q_{10} = ca. 2.0), reaches an optimum at about 30 to 35° C, and then drops rapidly to zero at temperatures around 40 to 50° C that inactivate the enzymes. At low light intensity, however, the photosynthetic rate rises very slowly with temperature (Q_{10} = ca. 1.1), goes over a broad optimum, and then drops down to zero at inactivating temperatures.

Some particular strains of algae that can tolerate unusually hot weather have been found to occur in hot springs and in Texas. The temperature necessary to inactivate their enzymes is higher than that of the more common algae. In these algae the rate of photosynthesis keeps on going up as the temperature is raised above the optimum for normal algae. As a consequence, much higher rates of photosynthesis than are usual can be produced at the temperature optimum of these special algae.

Elaborate mathematical attempts based on detailed theories have been made to describe the combined effects of various factors on the rate of photosynthesis; at one time this approach was thought capable of leading to a hypothetical mechanism that would account for the observed reaction. These attempts have, however, largely been abandoned in favor of the actual separation of chemical components and the study of separate steps in the reaction by biochemical isolation methods. The reason the straightforward kinetic approach used in the study of simple photochemical reactions was not successful when applied to photosynthesis is that too large a number of steps constitute the photosynthetic process. This does not mean that attempts to find kinetic explanations were wasted; it simply means that the results proved the actual mechanism to be too complicated for the kinetic approach to give a definitive picture of the system. . . .

The Pigments of Photosynthesis. Photochemistry is the study of the interaction of light and matter. A basic principle of photochemistry is that light must be absorbed in order to have any effect on a substance. Since light passing through an object cannot cause any chemical change in it, plants must have pigments if they are to use light. Pigments are colored because they absorb some colors more than others, and the reason for this selective absorption is that certain parts of their molecules have a natural vibration frequency corresponding to the wavelength of specific colors of light.

A quantum is the smallest unit package of light. Each quantum coming near a molecule is either entirely absorbed or moves on without being changed at all. If absorbed, it no longer exists as light and the energy of the light quantum is now within the absorbing molecule, giving it a higher energy content than that molecule had before the light absorption took place. The pigment molecule activated with this extra energy may do any one of the following things:

1) It may lose the extra energy by transfer to another molecule, thus leaving the second molecule activated.

2) It may cause a chemical change within the pigment itself.

3) It may waste the energy as heat and so speed up the random motion of the neighboring molecules.

4) It may re-emit a quantum of light as fluorescence. If this happens, the color of the emitted fluorescence will not be the same as that of the original quantum absorbed, but will be the specific color characteristic of the pigment molecule.

The possibility of a pigment molecule's performing any of these acts must be taken into account in the investigation of pigment function in plants.

The amount of light absorbed or reflected by a leaf depends on the color of the light. More of the light is reflected in the green part of the spectrum than in either the red or the blue. This is why leaves look green. Typical measurements of the reflection and absorption spectra of leaves are illustrated in Fig. 1.

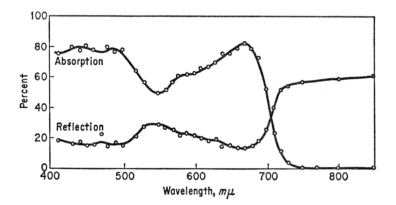

Fig. 1. The absorption and reflection of a cabbage leaf for different wavelengths of light. Leaves look green because they both transmit and reflect green light more than other colors.

It is evident from the curve that the reflection of leaves is high in the infrared and that there is almost no absorption at all by the leaf pigments beyond about 725 mμ. This high reflectivity of leaves in the infrared makes them look white in infrared photographs. The absorption peak in red light at a wavelength of 678 mμ is characteristic of chlorophyll in the special complex form that makes it able to do photosynthesis.

Light is essential not only for photosynthesis but also for the production of chlorophyll necessary to the process. The light that is used to form chlorophyll is absorbed by protochlorophyll. This is a pale green substance which needs only 2 hydrogen atoms to become chlorophyll but does not add them until it is illuminated. This reaction has been studied by J. H. C. Smith. Fig. 2 shows the effectiveness of different wavelengths of light in forming chlorophyll from protochlorophyll in illuminated corn seedings. The effectiveness is the reciprocal of the amount of incident light needed to turn a certain fraction of the protochlorophyll into chlorophyll. The wavelength position of the peaks very nearly matches the absorption peaks for

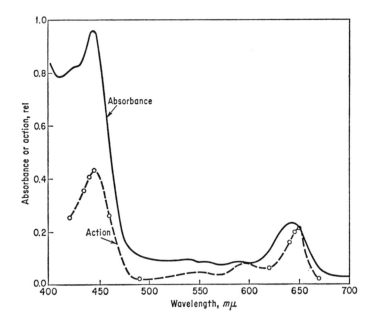

Fig. 2. The absorption spectrum of protochlorophyll in an etiolated albino leaf as compared to the effectiveness of different wavelengths of light in transforming protochlorophyll to chlorophyll. The agreement in wavelength of the peaks for the two curves shows that light absorbed by protochlorophyll causes its transformation to chlorophyll. (From J. H. C. Smith.)

protochlorophyll *in vivo.* This comparison shows that it is protochlorophyll which absorbs the light for its own transformation into chlorophyll.

Experiments have been made by means of growing plants in the dark and then exposing them to light until various amounts of chlorophyll are formed. When the leaves have only a very small amount of chlorophyll, the rate of photosynthesis increases directly in proportion to the amount of chlorophyll available. However, when an adequate supply of chlorophyll is available, an excess of it does not increase the capacity of leaves for photosynthesis. Some step other than the absorption of light by chlorophyll must therefore be limiting the rate of photosynthesis when plenty of chlorophyll is on hand. The photosynthetic system of plants has many other components in addition to chlorophyll. In fact, Smith has found that light is necessary for the development of some of these other, and as yet unknown, components of the cells' photosynthetic system.

It is possible to extract chlorophyll and the yellow carotenoid pigments from leaves by grinding them in alcohol. The different pigments in the alcohol solution can then be separated from each other. One of the ways to separate them is to pour an extract of pigments in a suitable solvent through a column of powder to which the pigments are adsorbed in varying degrees. Ordinary cane sugar is very good for this purpose if petroleum ether, with proper additions of other solvents, is used.

The pigments in solution can be separated from one another by pushing the column of powder out of the glass tube and cutting off various sections. The pigments thus purified can be redissolved and the absorption spectra of the pure pigments measured in a spectrophotometer.

By measuring the absorption spectrum of plant extracts it is possible to determine the actual amount of chlorophyll in leaves and to tell how much of it is chlorophyll *a* or chlorophyll *b*. This is done for an extract by comparing the height of the curve at the red peak with standard curves made with weighed amounts of the pure substance.

Fig. 3 gives the absorption spectra of purified chlorophylls *a* and *b* dissolved in ether. They have sharp absorption bands in the red part of the spectrum, another strong one in the blue, and various minor bands. Solutions of chlorophyll *a* are blue-green while those of chlorophyll *b* are olive green.

Chlorophyll *a* has the empirical formula $C_{55}H_{72}O_5N_4Mg$ and a molecular weight of 893.48. It is similar to hemin in that it has four pyrole rings, but chlorophyll has a phytol ($C_{20}H_{39}O_2$) tail. Chlorophyll *b* differs from chlorophyll *a* only

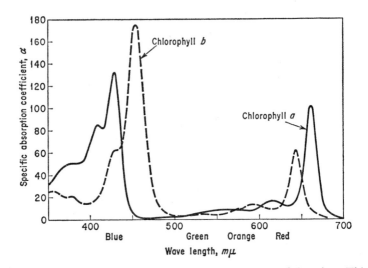

Fig. 3. The absorption spectra of chlorophylls *a* and *b* in ether. This curve was determined by Smith and Benetez by measuring the ratio of incident light, I_0, to transmitted light, I, of different wavelengths. Such reference curves are used to analyze plant extracts for their pigment content.

in having an aldehyde side chain in place of the one methyl group in chlorophyll *a*. Both chlorophylls are waxy solids insoluble in water but soluble in most organic solvents. In ether the red absorption peak used for the identification of chlorophyll *a* is at 662 mμ, whereas that of chlorophyll *b* is at 644 mμ. The reduction of chlorophyll brought about by light in certain solvents was recently discovered by Krasnovsky. This reaction has been suggested as being the way that chlorophyll might participate in photosynthesis. However, other ways in which it might act have also been suggested.

There are some similarities between the absorption spectrum of pure chlorophyll and of live leaves; at least the two spectra are enough alike to make it obvious that chlorophyll is what makes leaves green. However, the differences in spectra of chlorophyll *in vivo* and in extracts are very important. The wavelength shift from about 678 mμ in the leaf to

662 mμ in ether shows that extracted chlorophyll is a different chemical substance from the green material found in leaves.

Another difference between extracted and natural chlorophyll is its stability to light. A solution of chlorophyll in alcohol is bleached rapidly by strong sunlight in which leaves are able to thrive.

One of the major problems in biochemistry and biology, progress in which has lagged far behind many closely related scientific developments, is that of the chemical nature and mode of action of the chlorophyll complex found in leaves. It is extraordinary how little is known about the chemistry of natural chlorophyll, since the whole world of nature owes its green color to this substance. Nevertheless, its chemistry is as yet only very crudely understood, although a great deal is known about the properties of extracted chlorophyll. Chlorophyll in the leaf is thought to be combined with other substances, presumably proteins. It is known that the natural chlorophyll complex always contains carotenoids, colorless fatty materials, and protein, in large insoluble aggregates. Disintegrated chloroplasts, in fact, still consist of such large pieces of insoluble material that they make chemical analysis of the natural chlorophyll complex difficult and not particularly fruitful.

Analogies have sometimes been drawn between the hemoglobin of blood and the chlorophyll of leaves. However, there is far too much chlorophyll in relation to protein for the complex to be a simple protein-pigment compound of definite structure like hemoglobin, which has four hemins neatly attached to one protein molecule.

Chlorophyll in plants is attached to solid particles and its molecules are probably arranged in a thin layer on a protein surface. Chlorophyll therefore seems to have some of the properties of crystalline solids. It is believed to behave like a two-dimensional crystal — that is, a slice of crystal one molecule thick. Theories of solid-state physics that apply to electrical conductivity in crystals also describe the effects of light on the electrical conductivity of these chloroplast films.

Further complicating the study of chlorophyll *in vivo* is the fact that there are at least three different kinds of the chlorophyll *a* complex in plants. These forms of chlorophyll *a* have their light-absorption peaks at different wavelengths, and by these may be identified. The known forms of chlorophyll *a in vivo* are called C_a673, C_a683, and C_a695, and other forms may eventually be discovered. The absorption bands of these forms are about 15 to 20 mμ wide at their half height, so that they overlap and appear to fuse, giving the appearance of a single broad band.

In order to detect the presence of the different chlorophyll *a* forms it is necessary to measure the slight variations in shape of the composite absorption band. This can be done with a special kind of spectrophotometer constructed for the purpose of plotting the slope or the rate of change with wavelength of the absorbance curve rather than its height. These measurements of the first derivative of absorbance with respect to wavelength — particularly if made at very low temperature — show that bands of individual chlorophyll *a* components vary in their relative proportions in different plants, and even in the same plant when grown at different temperatures and light intensities.

The photosynthetic purple bacteria contain a different pigment: bacteriochlorophyll, which has its main absorption peak in ether at 770 mμ, a much longer wavelength than that of chlorophyll *a*. Like chlorophyll *a* in the leaf, the natural complex of bacteriochlorophyll in the bacteria also occurs in several forms. The difference in wavelength between the forms is, however, much greater and was identified by Wassink and others long before the different forms of chlorophyll *a* in plants were known for certain. The photosynthetic bacteria have chlorophyll complexes absorbing at 800, 850, and 900 mμ while those of chlorophyll *a* are at 673, 683, and 695 mμ.

The function of chlorophyll in photosynthesis is obviously to absorb light so that it can be used for producing chemical changes. Just how chlorophyll uses its absorbed energy is still the central question of photosynthesis. Does chlorophyll go through a reversible reduction of the Krasnovsky type? Does it transfer its absorbed energy directly to cytochrome? Does the natural chlorophyll complex act like a solar battery, separately + and − charges, thus driving oxidation and reduction reactions in different places?

Whatever the chemical state of chlorophyll in living leaves may turn out to be, it is certainly true that the chlorophyll molecules are packed tightly together. They appear to act in large units of several hundred chlorophyll molecules, so arranged that light absorbed by any one molecule can transfer its energy at random to another chlorophyll molecule within a single unit. Eventually this energy reaches an active center, possibly a place where a chlorophyll molecule is attached to an enzyme. This process of energy transfer is very important because it enables chlorophyll to use light with a high degree of efficiency. That is, light absorbed by any chlorophyll molecule is transferred directly to the spot where it is needed, instead of being wasted as heat if it happens to be absorbed by a molecule which is not attached to an enzyme.

This energy transfer not only takes place between chlorophyll molecules, but energy may also be transmitted from other pigments to chlorophyll. Some of the accessory pigments may accomplish this by transferring their absorbed light energy to chlorophyll, probably C_a673.

Chloroplasts of green plants contain two cytochromes, f and b, that are not found elsewhere. It seems likely that these cytochromes are closely bound with chlorophyll. Recent work of Chance has shown that even at extremely low temperatures the absorption spectrum of these cytochromes changes immediately when the chlorophyll is illuminated. It may be that the first chemical change in photosynthesis produced by light is a change in the oxidation level of these cytochromes effected by energy transferred from chlorophyll.

Not all the color in plants is due to chlorophyll; all photosynthetic plants contain some carotenoid pigments as well. Carotenoids are unsaturated hydrocarbons, though some of them may contain a few oxygen molecules. The name for this class of pigments is taken from the common carrot. Carrots are highly colored by beta-carotene, a pigment also present in leaves.

Red and blue-green algae, besides having chlorophyll *a* — though not *b* — and carotenoids, are often spectacularly colored by mixtures of the red pigment, phycoerythrin, and the blue pigment, phycocyanin. There are a number of different phycoerythrins and phycocyanins which as a class are called phycobilin proteins. The phycobilin pigments, like chlorophyll, are found as complexes with proteins and are also, like them, made of four pyrole groups. However, these phycobilin protein complexes are unlike chlorophyll in that they are water soluble, and their pigmented components cannot be separated from the protein as easily as chlorophyll is extracted from its protein complex. The absorption bands of various plant pigments are shown in Fig. 4.

Given a specific plant, it is possible to find out which pigments are participating in photosynthesis. This information is obtained by comparing the action spectrum of photosynthesis in the plant with the absorption spectra of the different pigments that are present. An action spectrum is determined by plotting the photosynthesis produced by a given number of incident quanta against the wavelength of light used.

Action spectra made by Engelmann in 1887 showed that the phycobilins and some carotenoids can use the light they absorb for photosynthesis. Anthocyanin pigments — those water-soluble glucosides that color red cabbage, beets, copper beech leaves, and many flowers — are not active in photosynthesis. All photosynthetic plant pigments other than chloro-

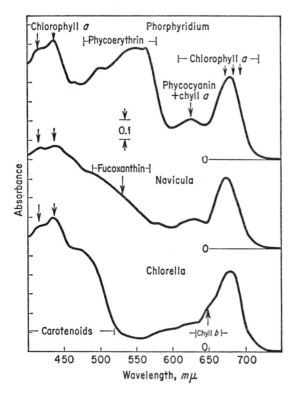

Fig. 4. The absorption spectra of three algae, showing the regions of absorption due to the different kinds of photosynthetic pigments. The large band at 675–680 is due to chlorophyll *a* in all the species. Three arrows show the positions of the separate forms of chlorophyll *a*, C_a673, C_a683, and C_a695 that overlap to make this composite absorption band. Chlorophyll *a* also has a small band around 620 mμ and two large bands in the blue region, marked with arrows. Nothing is known about the contribution of the different forms of chlorophyll *a* to these blue bands. *Chlorella vulgaris* is a green alga that has much the same pigment system as ordinary leaves. Chlorophyll *b* shows as a shoulder in the *Chlorella* spectrum and its blue band also causes part of the absorption in the carotenoid region at about 480 mμ. *Navicula minima*, a diatom, shows the absorption of the carotenoid fucoxanthin, which is at longer wavelengths than most other carotenoids. *Porphyridium cruentum* is a red salt-water alga with phycoerythrin and a very little phycocyanin.

phyll have come to be known as "accessory pigments" — a name with unfortunate connotations that for a long time helped to obscure the very important function of these colored substances.

Quite apart from the unsolved problem of how chlorophyll does its work we may ask whether the accessory pigments serve the same purpose as chlorophyll or whether they have an entirely different function. Within the last few years it has become evident that the accessory pigments do indeed have a specific role different from that of chlorophyll a. Furthermore, chlorophyll b and the shortest wavelength form of chlorophyll a, C_a673, have been found to have the same function as the accessory pigments.

Photosynthesis is now believed to depend on two different photochemical reactions driven by different pigments. Just what the chemistry of the two reactions may be is not clear, but there are biochemical reasons favoring the idea that one of them might be photophosphorylation and the other the formation of a reducing substance.

In green plants one of the reactions is driven specifically by chlorophyll b, by carotenoids, or by the 673 mμ form of chlorophyll a. The other reaction is powered by the 683 mμ form of chlorophyll a. In red algae it seems that whereas phycoerythrin and phycocyanin drive one of the reactions, all the chlorophyll a forms drive the other.

The existence of two separate photochemical reactions is deduced from the striking increase in photosynthesis obtained when two pigments both absorb light as contrasted with illumination of either pigment alone. This enhancement effect of two colors of light, each chosen to be absorbed by a specific pigment, was discovered by Emerson; it is at present one of the aspects of photosynthesis attracting much attention from a number of laboratories.

Objectives of Photosynthesis Research. There is a distant hope that with a more detailed understanding of how photosynthesis is carried on by plants it may be possible to increase their growth efficiency. This thought is at the back of the minds of all investigators in photosynthesis, but there is still no obvious means by which the data accumulating in photosynthesis laboratories can be applied directly to crop production. At present our understanding of photosynthesis is a part of the body of science that is almost pure and unapplied, and like all basic science has potential future applications, but in ways that are highly unpredictable. It is therefore premature to say that the reason for carrying on photosynthesis research is that the knowledge gained thereby will be of direct value to agriculture or to power production.

Understanding, as an objective in itself, has frequently become useful later from a practical point of view. Therefore it is clearly unnecessary to hunt for long-range applications to justify fundamental research on any significant scientific problem. Scientific understanding of natural processes has to be well developed long before the information can be applied to man's immediate needs. Certainly one thing that should be known about life on earth is how the food for living organisms is manufactured.

The history of photosynthesis well illustrates one important principle of scientific progress — namely, the limitation that incomplete knowledge in surrounding fields imposes on the possible development of any particular subject. In 1804, for instance, Saussure had clarified the nature of the photosynthetic process; he showed that carbon dioxide was used up, oxygen was produced, light was necessary, and organic matter was synthesized. The point is, that by 1804 the problem of photosynthesis had been solved — that is, solved in terms of the chemistry and physics known at the time. In 1845, the problem of photosynthesis was solved again by showing that it is really a conversion of energy in the form of light into a form of bound chemical energy. This transformation-of-energy aspect of photosynthesis could not even have been discussed before 1845 because the nature of the different forms of energy was not yet known.

A frequently raised question is: How long will it take to solve the problem of photosynthesis? To answer this question we must be clear as to what we mean by solving a problem. It is very likely that many of the questions we can now ask about how photosynthesis works will be answered within relatively few years. However, at such time as these answers are at hand it will also be possible to ask far more penetrating and detailed questions that cannot now be formulated. The continual expansion in the scope of questions that may be raised is one of the more important aspects controlling the advancement of any particular branch of science. In one sense no scientific problem has ever been completely solved, and in another sense, in terms of the state of knowledge at the time and the type of solution that would then have had meaning, most scientific questions have found their answer again and again — but each time within the limits of contemporary knowledge and vocabulary.

*D*r. Samuel R. M. Reynolds was born in Swarthmore, Pennsylvania, on December 9, 1903. He attended Swarthmore College, where he received the A.B. degree in 1927, the M.A. in 1928, and the honorary degree of Doctor of Science in 1950. He received the Ph.D. degree from the University of Pennsylvania in 1931.

He was a Fellow in medicine of the National Research Council at the Johns Hopkins University Medical School, Baltimore, in 1931-1932, and in the same year he first became connected with the Carnegie Institution, as Carnegie Fellow at the Department of Embryology. In 1932 he was instructor in physiology at Western Reserve University, and in 1933 transferred to Long Island Medical College, where he remained until 1941, becoming assistant professor in 1934 and associate professor in 1938. In 1937–1938 he studied with Dr. George Corner at the University of Rochester School of Medicine and Dentistry as a Guggenheim Fellow.

In 1941 he rejoined the Carnegie Institution, as a physiologist in the Department of Embryology. In 1950 his Guggenheim Fellowship was renewed for a year of study at Oxford University. In 1952–1953 he served as Acting Director of the Department of Embryology. While at the Institution he was the author of three highly significant books in the field of embryology: a rewritten, enlarged edition of his Physiology of the Uterus (1949); Physiological Bases of Gynecology and Obstetrics (1953); and Measurement of Uterine Forces in Pregnancy and Labor (1954). In 1956 he left Carnegie to join the faculty of the University of Illinois as head of the Anatomy Department, College of Medicine.

In addition to his academic career, Dr. Reynolds served as aviation physiologist with the U.S. Army Air Corps, and as member of the staff of the School of Aviation Medicine at Randolph Field and elsewhere from 1942 until 1945. He was visiting professor, faculty of medicine, University of Montevideo, Uruguay, in 1950, 1955, and 1956, and at Recife, Brazil, in 1955. He received the degree of Doctor honoris causa from the University Catolica de Chile in 1950, and was named professor honoris causa by the faculty of medicine at Montevideo.

180

Samuel R. M. Reynolds

THE UMBILICAL CORD

From *Scientific American,* July 1952.

Until the moment of birth a developing baby is entirely dependent upon the placenta. This vital organ, lying outside the fetus' own body, serves it as lungs, intestines and kidneys. From the placenta the baby gets its oxygen and predigested food, and to it it sends its wastes. The connecting link that carries this two-way traffic is the umbilical cord. From the time when the embryo is considerably smaller than half a small pea until the baby is born, the umbilical cord is its lifeline.

One might suppose that the functioning of so vital a structure would have been thoroughly investigated, but the fact is that it has interested relatively few people. Recently, however, we have learned some surprising things about the way in which blood flows through the umbilical cord and about the nature of the cord itself.

The fetus operates its own circulatory system. Its tiny, rapidly beating heart pumps blood through two arteries in the umbilical cord to the placenta. There the blood picks up water, sodium, phosphorus, iron, oxygen and other vital substances and flows back to the fetus through a single vein in the cord. Within a few months the blood must transport through the cord enough material to build a six- to eight-pound baby, and of course the total bulk of fluid exchanged between the placenta and the fetus is vastly greater.

The embryologist Louis B. Flexner and his associates at the Carnegie Institution have made some measurements of

181

this circulation. At the age of 14 weeks the human fetus weighs about two ounces and grows at the rate of about a quarter of an ounce per day. To gain at this rate the two-ounce fetus must daily "drink" and discharge (*via* the blood) about six quarts of water alone. At 31 weeks, when the fetus weighs three pounds, the cord must carry nearly 70 quarts of water per day. And water is only part of the whole blood volume. At 31 weeks the tiny fetal heart pumps roughly a fifth of a quart of blood per minute (nearly 300 quarts a day). The blood flow through the umbilical cord is rapid—it travels at an estimated rate of seven inches per second, about four miles per hour.

To carry such a load the umbilical cord must indeed be a remarkable organ. Let us examine its structure. The first striking fact is the wide variation in length of the cord. Its average length at birth is about two feet: that is, the cord is generally longer than the baby itself. But it may vary from as little as five inches to well over four feet. A physician in Valparaiso, Chile, recently reported a case in which the cord at birth was four feet eight inches long. It looped once around the baby's body, then over the shoulder, under an armpit and twice around the neck, with a good length left over to its root in the placenta. Twisting, twining and even loose knotting of the cord is the rule rather than the exception when it is average or above average in length.

About the cord's thickness it is difficult to be precise, because it cannot usually be measured except at the baby's birth, when blood is not flowing through it normally. Indeed, the fact that the cord is generally seen only after birth, when its blood vessels are collapsed, has led to a widespread misconception about its structure. Textbook pictures of the cord almost always show the blood vessels surrounded by a thick blanket of soft material called Wharton's Jelly, which is commonly supposed to serve as a cushion protecting the vessels. The picture is wrong: in the cord's normal state, when the vessels are distended by blood flowing through them, the jelly is not a cushion but a thin, tautly drawn band of tissue. We have demonstrated this by studies of a distended cord obtained in a Caesarean operation at Johns Hopkins Hospital.

That is really the beginning of the story which I have to tell about the characteristics of blood flow in the umbilical cord. When the jelly is stretched taut by the pressure of blood flowing through the vein, it acts as a tight membrane which is trying to collapse the vein at all times. The blood in the vein must be under appreciable pressure to keep the channel open for the abundant flow which is essential for fetal life. This pressure makes the umbilical cord a semi-rigid structure—an erectile type of organ. Why should this be necessary? And how is it achieved?

One afternoon in the winter of 1948 the Harvard University obstetrician Seymour Romney came to the Carnegie laboratory to discuss problems of blood flow in the placenta. He brought with him a number of models prepared by filling the blood vessels with a latex compound and then dissolving the tissues. What remained was a rubber "skeleton" of the lumens (bores) of all the larger blood vessels in the placenta. All of us who saw these preparations that day were struck at once by the fact that at the place where the blood vessels come into the placenta from the umbilical cord and branch out, the arteries and veins were about equal in number and of roughly the same diameters. This must mean that blood flowed in the veins at approximately the same speed as in the arteries. That seemed strange, considering that in the veins the blood is under much lower pressure. Whence came the energy to drive the blood through the veins at such high velocity?

We decided to investigate the situation in the umbilical cord, which is easily accessible and contains only three major blood vessels—two arteries and a single vein. From a number of observations in experimental animals and a few human cases, we calculated that the cross-sectional area of the two arteries combined is smaller than the cross-sectional area of the vein, and that the velocity of blood flow in the umbilical vein is between 70 and 75 per cent of that in the two umbilical arteries.

At this point we called in two hydraulic engineers and a mathematician at Johns Hopkins University to help up look into the mechanism responsible for the rapid flow in the vein. G. F. Wislicenus, a noted specialist in hydraulic engineering, was at once impressed by the fact that the umbilical vein is distended under considerably more pressure than ordinary veins are. The pressure in the umbilical vein appears to be at least 20 to 25 millimeters of mercury, which is about 10 times the pressure in the large veins near the heart in a normal adult. It is about half as great as the mean pressure in the umbilical artery.

When Professor Wislicenus learned of the unusually high pressure in the vein, he asked:

"How does the umbilical vein end in the baby?"

A rough pencil sketch was drawn. Blood flows from the umbilical vein to the fetal heart by two pathways. One is a roundabout route through numerous small branches into the liver and thence through the hepatic veins. The other route is much more direct: from the umbilical vein a special duct called the *ductus venosus* leads into the inferior vena cava and so to the heart. Beyond the duct blood flows rapidly but at the low pressure of one to two millimeters of mercury.

As soon as Professor Wislicenus saw the sketch, he put his pencil on the *ductus venosus* and asked:

"Where is the valve here that controls the pressure in the vein?"

"There is no such valve," we replied.

"Oh yes, there must be if the pressure in the vein is as you say it is. If there were no valve, the blood would flow in the path of least resistance; it would take the direct way to the heart and avoid entirely the resistance offered by the liver."

We went back to our laboratory, hunted through a

new book on *The Foetal Circulation* just published by A. E. Barclay, M. M. L. Prichard and K. J. Franklin of Oxford University, and within a few minutes found the very structure Professor Wislicenus had postulated. There were even X-ray moving pictures of the *ductus venosus* closing under certain conditions. This closure had been thought to be due simply to a sphincterlike contraction of the vessel to block back flow after the umbilical cord was cut and prevent undue loss of blood; no one had ever suggested a function for this structure during the life of the fetus. When Wislicenus said that it was a valve which served to regulate the pressure in the umbilical vein, he was asked what sort of a hydrodynamic mechanism might be expected to operate under these conditions.

"It may be that as the arteries are distended by pulsating pressure, they press upon the vein, which lies beside them in the same sheath. Since a floodhead of blood is pushing into the vein at the placental end, the arteries' pressure on the distended vein should force blood toward the baby end, where the pressure is lower. In other words, it is a supercharged system and appears to be acting like a pistonless pump."

This was the beginning of a new conception of the mechanism of venous return from the placenta. Like most early concepts, it required modification in the light of later observations. But it opened our eyes to new avenues of exploration.

The mathematician we consulted was F. W. Light, Jr., a former physician who was teaching higher mathematics at Johns Hopkins. He had maintained an interest in the mechanism of pulsating arterial flow, and had long concerned himself with the idea of expressing mathematically the characteristics of such flow.

Light asked three questions: (1) What are the pressure gradients along the umbilical arteries and vein? (2) What changes in diameter occur along the artery as the pulsation due to the pumping of the heart rises and falls? (3) What are the velocity and quantity of blood flow in the arteries and vein of the umbilical cord? Not only did Light want each of these bits of information; he wanted them from the same vessels at one and the same time and without cutting or entering them.

How could the necessary data be obtained? The answer seemed to lie in the studies on fetal circulation by the Oxford workers. This group, working at the Nuffield Institute for Medical Research, had obtained numerous superb X-ray photographs of the blood flow at frequencies of three to six pictures a second. What could be simpler, I thought, than to place a sheep fetus with the umbilical cord still attached to the placenta over a camera, inject an opaque substance and get a moving picture of it passing through the cord's arteries or vein? The deformation of the arteries could be measured from the picture, the blood flow could be timed, and blood pressure could be determined by conventional techniques.

With this end in view, arrangements were made for me to spend nearly a year at the Nuffield Institute, working with the enthusiastic cooperation of a group of investigators there, especially Prichard of the Barclay team, the radiologist G. M. Ardran and G. S. Dawes, the director of the Institute. The facilities at the Institute and the experience of its workers were uniquely fitted for investigation of the questions Light had asked. In particular, it had excellent equipment for making X-ray movies.

To measure the pressure gradients in the umbilical arteries while blood was flowing through them, we had to develop a specially adapted technique, for we found that the usual method distorted the pressure. A popular method is to insert into the blood vessel a small hypodermic needle filled with a salt solution and connected to a sensitive electrical pressure-recording device. But we found that the insertion of the needle into the artery deformed the wall of the artery and distorted the blood flow. Instead, we inserted as large a needle as possible into the first branch of the umbilical artery and advanced its blunt tip to within about one-half millimeter of the true wall of the main artery. This yielded an accurate measure of the lateral pressure in the umbilical artery as the blood went rushing by the branch on its way to the placenta. We had not interfered with more than a very small part of the total outflow of the artery, and we did not touch the pattern of stream lines at all. At the same time, the cord was placed over a specially built camera which moved 5-by-7-inch films directly beneath the entire cord at the rate of two per second. By running through 24 plates and having them synchronized with a known phase of a single pulse pressure-wave at the start, it was possible to determine exactly when in the course of succeeding pulse-waves a given picture was taken.

This experiment answered two of Light's questions. First, we found that the diameter of the artery is the same throughout its length. Since the artery is trying to collapse all the while, it is clear that the blood pressure also must be essentially the same along its whole length. This was a surprising finding, because one would assume that friction between the flowing blood and the sides of the vessel would progressively reduce the pressure and hence the diameter of the vessel as the blood moved farther from the heart. There must be some pressure gradient, of course, but it was not sufficient to affect the size of the vessel.

The answer to the second question was even more surprising. We could find no evidence that the periodic pulsation of the blood pumped by the beating heart produces any significant expansion of the artery. To anyone who has ever put his finger on the pulse and felt the pulse-beat pushing on his finger, this must seem incredible. Yet in excellent motion-pictures taken at 25 frames per second and in frame-by-frame comparisons of the artery's diameter we were unable to detect any broadening of the artery whatsoever as the pulse-wave passed along the vessel. Furthermore, we found that this situation is not peculiar to the arteries in the umbilical cord. We made motion-picture studies of the aorta of a cat. The section of the aorta immediately next to the heart did swell as the heart pumped blood into it. But the load was quickly dissipated, and along the rest of the aorta there was no measurable increase in the vessel's diameter.

What is it, then, that the doctor feels when he takes your pulse? We can only guess, but we have one significant clue. Sometimes an artery can be seen to move with the pulsebeat. This occurs only when the artery is curved, and it is due to the fact that the pulsating bloodstream pushes intermittently on the curved wall of the artery, moving it sidewise. This gives it a kind of lashing motion. When one presses on the pulse with a finger, he bends the artery into a curve, and the pulsation he feels is the beat of the pulsating stream against the deformed part of the vessel.

The third point that concerned Light—the velocity of blood flow—is easily determined by injecting a radiopaque substance into the bloodstream and measuring the rate of travel of its shadow on a film. Since the diameter of the vessels may also be measured from the shadow cast in the same pictures, it is possible to calculate the quantity of blood carried in the arteries or vein in a given tme. From such measurements a number of useful and significant facts have been learned.

In the first place, the quantity of blood flowing through the two umbilical arteries is, within certain limits, independent of the diameters of the arteries. It is governed, rather, by the needs of the fetus. The flow is of the order of 300 to 400 cubic centimeters of blood per minute per kilogram of body weight of the fetus. This amount, which goes to the placenta, is approximately two-thirds of the total output of the fetus' heart. It is striking testimony to the importance of the placenta as the seat of fetal nutrition and elimination.

In the second place, the volume of blood flow gives us a means of figuring out the pressure gradient in the umbilical arteries. Although there is no visible evidence of a fall of pressure, there must be some fall due to the frictional resistance of the sides of the vessel. From the known volume of flow, the radius of the artery and other physical constants of the system, we have calculated that the pressure drop is of the order of one-half to one millimeter of mercury for each centimeter of length of the vessel; thus in a cord 10 centimeters long the total fall of pressure is equal to 5 to 10 millimeters of mercury.

Now this is a very small frictional loss of energy. The frictional resistance of the vessels is so small that the blood will pour from the placenta to the fetus by virtue of even a small difference of hydrostatic pressure between them. Since the larger a vessel is, the less the proportionate friction along the walls, we can see now the purpose of the relatively high blood pressure in the umbilical vein: it keeps the vein distended to minimize friction.

The mechanism by which this pressure is maintained and the blood is kept flowing rapidly remains to be worked out. Evidently Wislicenus' original idea of the "pulsameter" pump, with the pulsations of the artery pressing on the vein, is not quite correct. For one thing, we have seen that the artery does not swell as it pulsates, and for another, we have found that in an animal such as the sheep the arteries in the umbilical cord are not even adjacent to the vein but are separated from it by an appreciable layer of jelly.

How might a pulsameter pump operate in this situation? We measured the pressures simultaneously in the umbilical artery and in the umbilical vein. It was found that the pulse pressure-wave in the umbilical vein was almost exactly 180 degrees out of phase with that in the artery! The pressure in the vein decreased as that in the artery rose, and *vice versa*. Here was evidence of a pulsameter pump, but the time relations were exactly the opposite of what had been originally suspected, for if the artery pulsation pressed on the vein one would expect the pressure in both vessels to rise simultaneously. The only explanation that can be offered at present to account for the actual state of affairs is that the artery increases slightly in length during each pulsation. But we do not know that this is so. . . .

*F*orm, Forces, and Function in Plants," Dr. Herman A. Spoehr's *felicitous discussion of plant biology, appeared originally in* Cooperation in Research, *a publication of the Carnegie Institution. It was later published under the title* Essays on Science *by Stanford University Press in 1956. It is presented here in somewhat abridged form.*

Herman Spoehr was born in Chicago in 1885, and majored in chemistry at the University of Chicago. After his graduation in 1906 he traveled to Europe, where he studied carbohydrate chemistry in Berlin, and at the Sorbonne in Paris. The following year he returned to the University of Chicago to do his Doctoral research in chemistry and to study plant physiology.

He joined the Carnegie Institution in 1910. His first ten years with the Institution were spent at the Desert Laboratory near Tucson, in what in 1910 was the Territory of Arizona. Appropriately enough, Spoehr carried on his pioneering research on the chemistry and physiology of plants under frontier conditions. He rode over the desert to the Laboratory on the Territory's first Harley-Davidson motorcycle.

In 1920 Spoehr moved to the Carnegie Institution's Coastal Laboratory on Carmel Bay, California, where he continued his research on carbohydrate chemistry and photosynthesis, and wrote Photosynthesis, *a book called by one of his colleagues "the bible for a generation of workers in this field." With recognition came greater extra-scientific responsibilities. In 1928 he was named Chairman of the newly organized Carnegie Institution Division of Plant Biology, with headquarters on the Stanford University campus.*

He left the Carnegie Institution in 1930 to serve as Director of the Natural Sciences Division of the Rockefeller Foundation. Returning in 1931 to the Institution, he resumed the Chairmanship of the Division of Plant Biology, and also found time to continue his studies on the composition of complex carbohydrates in leaves, the conditions required for the activity of leaf enzymes, and the large-scale culture of algae. In World War II he helped develop methods of extracting antibiotics from Chlorella algae.

He resigned his administrative duties at Carnegie in 1947 to devote more time to research, which he continued even after retiring in 1950. He also served as the first Science Adviser to the Secretary of State in 1950–1951. He died at Palo Alto, California, in 1954.

190

H. A. Spoehr

FORM, FORCES, AND FUNCTION IN PLANTS

From *Cooperation in Research,* Carnegie Institution, 1938.

In the old university town of Uppsala, some two hundred years ago, there was hatched a scheme which for daring and scope had rarely been equaled. It was one of those presumptuous, Utopian plans which evolve only in the unfettered mind of youth, a plan demanding more imagination and of vastly greater magnitude than any conceived by Gustavus Adolphus or Napoleon. In many an older head similar aspirations had been cherished, but most of these had been quenched by the very magnitude of the undertaking. It was given to two young, impoverished students to conceive and shape the principles of the scheme. As a first step toward its realization they divided between them the three great kingdoms of the earth — the mineral, the vegetable, and the animal. Their plan was to order, arrange, and name all there was, and all there was to be, in these great domains, according to their *systema.*

These zealous youths were Carl Linnaeus and Pehr Artedi, by paternal design respectively students of medicine and of divinity. But fate overrode parental ambitions; Artedi was drowned in a Dutch canal, and it was left to Linnaeus, though he finished his medical studies, to carry out and elaborate the great plan.

Observant men had long been impressed with the tremendous diversity of nature, particularly among the living

191

things. The multiplicity of form taken on by living organisms seemed to present an almost hopeless wilderness to the orderly-minded. The great voyages of exploration had increased these domains apparently beyond any hope of survey; from the Americas, Asia, Africa, from the teeming tropics and the sparse deserts new and strange creations appeared. There never was a period when the need for order and the urge to arrange were so perfectly timed. Out of these efforts grew a universal language, the elements of which had been envisioned by Linnaeus and Artedi. And the magnificent feature of the scheme of these two youths was not only that it was, in fact, set into motion and pursued by Linnaeus during his entire life, but unlike all the conquests of the Napoleons, which so rapidly fell into ruin, this youthful endeavor prospered, and grew with altering needs long after its guiding spirit had gone to his rest.

Looking back over two centuries of restless activity of this army of collectors, arrangers, and classifiers, of this cease-less effort to discover some thread which binds like to like amid the tangle of multiformity, one is impressed not only with the enormous industry and persistent application shown, but even more with the stupendous compass of the domains concerned. The urge to systematize and to order penetrated all substances and forces of the earth. From this effort sprang separate disciplines devoted to particular fields or kinds of substance. Because of limited diversity some fields were more rapidly arranged and their tillers could turn from classification to measurement, from mere ordering to the determination of the more exact arrangement of the parts. "Classification is a half-way house between the immediate concreteness of the individual thing and the complete abstraction of mathematical notions."* Thus much of early inorganic chemistry was devoted to efforts at classification, with the great reward of the periodic system and its amplifications. Out of the early efforts at forming types, organic chemistry rose to be an exact science.

---

* A. N. Whitehead, *Science and the Modern World,* Cambridge University Press, 1928.

But these were relatively simple tasks as compared with the classification of living beings. The number of "species" or units of unorganized matter was relatively small, and, moreover, it could all be done within the comfortable confines of a laboratory. The living organisms offered far greater diversity and had to be brought in from the far corners of the earth, from the air, from the waters, and from the waters under the earth.

In this short period of time, short as intellectual efforts must be measured, there has been wrought at least a definite semblance of order. And most of this has been accomplished since Linnaeus set out on his great venture. No one would be so rash as to claim finality for any of these arrangements. But they have given us a remarkable working basis for the further steps. Here we have the count of this task, naturally in approximate figures:

92 elements
1,500 minerals
30,000 inorganic compounds
300,000 carbon compounds
1,500,000 species of animals and plants

It stands as an eloquent expression of a faith, an intuitive faith that there *is* order in nature, which is the impelling motive of science. This faith has been rewarded with at least a measure of understanding and, stimulated by vivid imagination, has sustained the ever-quickening labors to penetrate the unknown.

The tremendous diversity of living beings and the fact that many of these were confined to particular and often inaccessible parts of the world, together with their complexity of form and function, imposed upon the study of plants and animals a much heavier burden than was the case in the efforts to classify the elements, the inorganic compounds, and even the carbon compounds. Yet this very diversity of organic nature seems to have offered opportunity for the discovery of fundamental phenomena not so easily detected in the simpler systems. Ideas of the fixity and permanence of the elements grew with the advancement of the science of chemistry, and the immutability of these units was long regarded as an established fact. The lack of permanency in nature and the concept of mutability were recognized in biology long before they were in chemistry. Ideas of organic evolution penetrated the thinking of biologists and were considered as an important phenomenon in organic nature long before such concepts were advanced in chemistry through the discovery of radioactivity and of isotopes. But the discovery of transmutation of the elements was followed rapidly by an understanding, or at least a very acceptable theory, of the mechanism which was involved. On the other hand, biology can claim little more than that it has described evolution, and in a measure reconstructed the steps through which it has passed. The means by which these changes occur in the form of living things remains one of the greatest problems of science and of all humanity.

FORM

It is thus generally recognized that the diverse forms of animals and plants, as found in the world today, represent one stage in the history of organic evolution. The arrangement of these forms in such manner as to display their natural interrelationship and to aid in visualizing their stage of evolution has become an outgrowth of the early efforts at classification. Systematics, through the differentiation of organisms on the basis of their morphological characters, is one of the most time-honored methods of unraveling the apparently limitless variety of forms which are now found in nature, and of others of which we can discover only the lifeless remains. The theory of organic evolution has as its outstanding contribution the demonstration, so far as this is humanly possible, that these organisms have arisen gradually, and that many of the changes in their form have occurred as a result of influences which continue in operation and which are therefore discoverable. . . .

Fortunate the man who can go directly to nature for an answer to his questions. Early in the organization of the Carnegie Institution it was recognized that the western portion of the American continent offered unusual opportunities for studying life in its native environment. Great areas of virgin country had been made accessible by modern transportation, but had not yet been altered by the devastating hand of man. A tremendous range in climatic conditions is here reflected in a flora and a geographical distribution thereof which are extraordinarily varied. . . .

At several localities various problems relating to the influence of environment on form and distribution of plants have been intensively pursued, advantage being taken of the special environmental and vegetational characteristics of the particular region. Thus in the arid regions of the Southwest, in the Coast Range and Sierra Nevada of California, in the areas occupied by the great redwoods, and on the Great Plains, the influence of geographical and climatic factors was mapped, analyzed, and subjected to experimental study. Although the particular plant organisms and environmental complexes which herein were made the subject of special studies exhibited great variation, and the viewpoints and approaches of the particular investigators differed in many respects, it has become evident that the fundamental problems involved have many elements in common and are amenable to resolution on the basis of recognized scientific principles. . . .

Since the founding of the modern science of taxonomy by Linnaeus, living beings have been classified on the basis of the form of certain of their parts. In this, the comparison of the visible and also some of the more minute parts has been used as a means of differentiating and classifying organisms into categories.

Of the approximately one and a half million species of living beings which have been thus recognized, about 133,000 are flowering plants. The general classification of a great many of this number has been fairly well worked out, and the "species" has been generally accepted as the central unit. But many of these so-called species, upon more careful examination, show differences, polymorphous characters, which in themselves are perhaps not sufficiently significant to warrant further differentiation by name into other species. The species may thus be composed of further recognizable units constituting subspecies, varieties, races, biotypes, or combinations of ecological forms. These intermediate or weakly differentiated forms are the "small species" of refined taxonomy. They constitute the material for much of taxonomic controversy. But of far greater importance is the fact that they may be the ray of hope for the student of evolution. The careful study of such polymorphous groups of very wide distribution may lay the basis for an understanding of the changes which have occurred in the processes of evolution — of differentiation from the ancestral stock during geographical dispersal or migration, and through the lapse of time. . . .

Hybridization has been found to be a particularly useful experimental test to differentiate the array of forms constituting a polymorphous "species" such as is here being considered.

It has been concluded by the investigators conducting these researches that these forms (or groups of similar individuals) are separated by mechanisms which are inherent in their own structure. Thus true species are separated by genetic barriers or incompatibilities. Their hybrids are either partially or completely sterile according to the degree of incompatibility. In extreme cases species are so incompatible that no fertilization takes place; in others fertilization is possible, but the development of the first generation hybrid stops in the early embryonic stage. Others have seemingly good seed which, however, does not germinate; while still others yield hybrid seedlings which are dwarfish, slow-growing, or which die before maturing. Finally, some species can form vigorous and normally flowering hybrids in the first generation, but they are entirely or almost entirely sterile; or sterility appears in the next generation. The incompatibility is expressed in many ways and gradations, but in the first or succeeding generations abnormalities and loss of vigor in one manner or another becomes apparent, so that finally a very large percentage of the hybrid product will be wiped out because, from lack of vigor, it is unable to compete in nature. The ramifications of this manifestation of incompatibility in interspecific hybrids are very great and the detailed cytogenetic elucidation thereof constitutes in itself an intricate study. For the present purpose it must suffice to emphasize that species represent a stage of evolutionary process "at which the once actually or potentially interbreeding array of forms becomes segregated in two or more separate arrays which are physiologically incapable of interbreeding."†

By similar means it has been found that what are termed *subspecies* or *ecotypes* produce hybrids which are fertile and vigorous in both first and second generations. There are no apparent genetic incompatibilities. They appear to show geographic or ecologic preference, and the characters by means of which they are distinguished probably have selective value in their natural environments. Lacking genetic incompatibility, they keep pure only through their geographic or ecologic isolation. Where they meet in nature and cross, purity is probably attained by natural selection....

FORCES

One of the most significant features of plant life is that each individual is permanently bound to the environment in which it starts life. This is more particularly true of the higher, flowering plants. From the time of the germination of its seed till its death the plant's entire existence is spent in one spot; it is there exposed to all the elements and changes of season. It is self-sufficient in that it is capable of manufacturing its own food. If it can withstand the forces of weather, if it can

† T. Dobzhansky, *Genetics and the Origin of Species.* New York: Columbia University Press, 1937.

secure the necessary raw materials for its manifold functional activities, it will be able to survive. Between mere survival and luxuriant development are innumerable stages in all of which the forces of the environment impress themselves upon the development of the plant.

These forces of the environmental complex are constantly changing in relative intensity and effectiveness. No two hours, days, or seasons are an exact repetition of previous ones, no two spots on the earth are exactly the same so far as the plant is concerned, and with greater changes in location corresponding modifications in environment exist. Most plant migrations are effected by means of seed dissemination, by which means the offspring is always carried into an environment differing more or less from that of its parents. To some of the changes in the intensity of environmental forces the plant is indifferent; others may affect its life profoundly. Also some plants are extraordinarily sensitive to very slight modifications in their environment, while others are able to exist over a wide range. Examination of the diversity of form taken on by plant organisms clearly suggested that form was an expression of the forces in the environment to which the organism was exposed. More critical and minute study of variations in form as used in classification, where form is considered in terms of development, has given further indication of the influence of changing forces in course of time, and their resulting effects.

What applies to the form of plants in relation to the environment applies equally to their function; in fact the three—form, forces, and function—are so intrinsically integrated in the living world that no one of them can be adequately considered without regard for the others. Life is essentially a dynamic concept; the organism is characterized by functional organization, operating in an environment which, because of its inherent physical and chemical nature and because of the influences of organisms within it, is itself dynamic. The organism cannot be regarded simply in the light of a highly organized machine operating within a physical environment, or as a thing set apart from the inanimate world. It is too intimately a part of this environment, too intimately interlinked with the physical forces of the environment, to be so regarded if a comprehensive concept of nature is to be obtained. . . .

The desert. Among the first projects undertaken by the Carnegie Institution was one devoted to the investigation of life under extremely severe environmental conditions. The story of the emergence of life from the sea had been written, at least in outline; and while innumerable problems still remained and were, in fact, multiplying in hydra-headed fashion, it was realized that many novel lines of development were presented by land plants. In these there had been perhaps no more differentiation with reference to reproductive processes than occurred in marine organisms, yet the land plants showed vastly greater diversification in the vegetative or non-reproductive features. It was realized, moreover, that scarcely less striking than the attainment of the land by plant and animal life has been its gradual colonization of the drier and drier parts of the earth's surface. The movement of the biota from moist lands to dry ones or their adaptation to changing conditions in the same place has resulted in the development of striking structures in plants and profound changes in their life habits. These remarkable effects in plant life are most strikingly exhibited in the extraordinarily severe climatic conditions of the desert areas. Such areas, or semi-arid ones, constitute one-third of the earth's land surface and nearly one-fourth of the area of continental United States. In terms of physical environment such as extremes of temperature, of lack of water over long periods of time, of intense sunlight, parch‧ing winds, and cold, clear nights, the deserts present conditions which represent the extremes for the survival of life. Yet this delicate, sensitive stuff called protoplasm, which is the carrier of all life activities, has proved to be more stable and persistent than the dark mountains and the very rocks among which the plants of the desert grow.

The enigma of the desert is its life. Our notions of living things, their form, how they function and their relations to each other, etc., have for the most part been formulated from experience and observation in temperate or tropical regions. The contrasts which one first encounters on the desert are startling and harsh. But the true depth of these contrasts and their objective significance reveal themselves only gradually through intimate living in the environment. Whence came these strange plants and animals which inhabit the desert and how are they able to maintain their life, to adjust their functioning to the extreme physical conditions to which they are exposed? These are the recurring questions; for these organisms have retained the functions essential to all higher plants, such as the manufacture of their own food, growth, and reproduction, by apparently an astounding capacity to modify some of the ways of their ancestors and to become organized for the environment in which they find themselves.

It was essential, first of all, to obtain a picture of the vegetation of the arid regions on the background of the complex of environmental conditions of which it is a product. This has been done by means of extensive exploration and study of the composition, relationships, and distribution of the desert vegetation. It is now apparent that in many respects the plants of the desert are less profoundly influenced by one another than are those of moist regions, and that they are more closely affected by the environment. . . .

For a visualization of the features which are truly characteristic and significant it has seemed important to make comparative studies of other arid regions, and this has been done throughout North America, and in Algeria, Libya, Australia, and South Africa. Identical or closely similar features in structure, function, and adjustment to environment are found

again and again in far-separated regions or in different continents. These universal features are definitely related to the ways
in which the evolutionary processes have molded the vegetative organs of plants into successful desert forms. . . .

The desert's physical environment itself presents an
extremely complex picture, especially when this is regarded
in its possible relation to the life of the region. Not only do the
various climatic factors, such as rainfall, evaporation, and temperature, vary over a wide range, from those that are extremely
unfavorable for organisms to those that are merely difficult;
but the distribution of such climatic factors during the course
of a year, for example, is quite as important for survival as their
intensity. The long and often irregular periods during which
the plants are exposed to extremes impose rigors which are
definitely determinative for survival. A true conception of these
physical conditions can be gained only on the basis of extended
and painstaking measurements over wide areas. The accumulation of such records and their correlation with the life habits
of the organisms alone can form the true picture of the environmental influences. For example, the ratio of rainfall to evaporation and the manner in which water is held and supplied to
plants by soils have been found to be highly important in controlling plant distribution and in determining the limits of the
desert, the line between grassland and forest, and the range
of certain forest types.

The extensive experience, now extending over thirty
years, gained from the study of the relation of climate to
organism has clearly demonstrated the fundamental importance of time in these relationships. A single season or year,
or even a few years, of observation can yield only a fragmen-
tary picture, and often a quite erroneous one, of the interplay
of climatic factors on living organisms. Whether the variations
in climate are truly cyclical or whether these only approach
cycles in the astronomical sense, as well as the definite correla-
tions between solar and terrestrial phenomena, obviously will

require further study and elaboration. The fact is nevertheless clearly established that there are recurrences of climatic conditions favorable and unfavorable to the development of plants. Moreover, it is now established that the great mass of observations on the relationships of plant and animal life to these environmental conditions is definitely amenable to scientific collation. The concepts arising from such investigations will ultimately have real social significance and be of human service. They constitute the only rational basis for a solution of the problems which are presented in the management of the half-billion acres of arid and semiarid range land in the United States, and the larger areas in Africa, Australia, and South America. . . .

FUNCTION

The functioning of plants has been studied from quite a different viewpoint from that followed in the efforts at systematization. Instead of stressing the differences and peculiarities which distinguish different forms of plants, function has been regarded largely as a property which is characteristic of all plants. Especially is this true of the higher, green plants.

The most significant of such common functions is perhaps that all living beings, just as all man-made machines, require a source of energy to maintain their activities. For the running of an engine combustible fuel is required. The maintenance of the life activities of a plant or animal organism requires a supply of fuel in the form of food. The energy from the fuel for the engine is derived through the burning of the fuel, that is, by its chemical combination with the oxygen of the air. The energy from food is obtained by the organism through an analogous though more intricate process. But in general principle the analogy between organism and engine holds. In all this the green plant exhibits one very important difference: it can manufacture its own food. All animals are entirely dependent upon the environment for their supply of food; so also, of course, is man, and so also does he take from the environment the fuel which runs his machines. To the extent that the green plant alone is capable of manufacturing its own food, it is a far more independent and self-contained being than any animal. Yet the plant, in turn, is dependent upon its environment for the raw materials out of which it manufactures its food and for the energy which it uses in this process.

By means of this process of food manufacture through photosynthesis the plant lays by regularly a small reserve for use during periods of stress and adverse conditions. It apparently has definite regulatory devices which prevent the complete consumption of its daily earnings during fair weather. Survival of the individual plant depends upon an available supply of reserve food for adverse times; survival of the species depends upon passing on to the next generation a patrimony in the form of food material, to enable the young plant to pass successfully through the critical stages of its early life to the time when it is itself an independent photosynthetic organism. All animals, including man, owe their existence to robbing the plant of these reserves. These reserves of the plant, either directly, or indirectly through first having nourished an animal, constitute our sole means of subsistence. Similarly our industrial needs for fuel, both coal and oil, are met by drawing upon the remains of plants from past geological ages.

These rapidly vanishing supplies we regard as our most important natural resources. In fact they are but a speck, almost an accident, in the great interlinking of life and environment on this planet. The whole environment, animate and inanimate, is in fact inextricably involved in man's well-being. Not only the consumable resources are involved in this, but quite as much the end products or results of man's activity. Life cannot be regarded only from the viewpoint of what is required from the environment for life's maintenance; of equal significance, especially for future generations, is what this life is doing to the environment.

To digress for a moment with an illustration which may not be without some social implication: To the biologist the phenomenon of hardship in the midst of plenty is not uncommon, since he has learned to consider not only a single need of an organism, such as food, or the organism alone, but rather the entire system in which the organism is active and its reciprocal relation to all factors which may affect its life. For example, the minute yeast plant has the capacity of using sugar as food. It works rapidly and, with adequate food supply, prospers and multiplies. The products of its activity are alcohol and carbon dioxide. These accumulate in its immediate environment, and they are both harmful to life. If sufficient food is available, the organism continues its fermenting action, and the concentration of its metabolic products soon reaches a point which makes it difficult for the weaker members of the colony to live. With the continued activity of the more vigorous members, the metabolic products increase to a point which causes the death of a large proportion of the colony, until finally the entire community falls victim to the deadly influences of its own metabolic products. Thus, in spite of an ample supply of food and of the vigorous work of its members, the life of the community is lost, a victim of its own activity. The same phenomenon can be observed in organisms which form butyl alcohol from sugar; and many others, such as the lactic acid organisms, survive only when the products of their activity are removed by suitable means from the environment. Obviously the living organism may influence its environment, and consequently its own life, not only by taking materials from this environment, but quite as drastically by adding other materials to the environment which produce its own destruction, in spite of a surfeit of food.

In the first part of this article attention was directed to the tremendous diversity of organic nature. It is not without interest that the approximately one and a half million different forms of living beings, all of which show some differences in life habit, are dependent for their existence upon a single chemical process, namely that of photosynthesis in plants. As an over-all process it is a relatively simple chemical reaction: carbon dioxide and water, under the influence of light, yield a carbohydrate, with the liberation of oxygen. This is the mother reaction for almost all life. Through all the great diversity of plant life, representing widely divergent evolutionary development, there is no evidence that this means of obtaining food has in any way been modified. With the exception of some of the colored bacteria, and possibly some of the algae, all plants apparently follow the same course of obtaining their primary food. Within the body of the plants this primary food is changed into a great multiplicity of chemical substances. Herein the diversity of plants again comes to expression, for the variety of substances found in plants is enormous and differs greatly with the life habit and environment; differs, in fact, from species to species. Within the bodies of animals these carbon compounds are again altered, some broken down to simpler ones, others built up into more complex forms. The three hundred thousand carbon compounds, mentioned in our first table, include the host of substances found in plants and animals. Besides these, a very large number of carbon compounds has been synthesized in the laboratory. The great majority of these are directly or indirectly connected with the substances formed in the photosynthetic reaction of plants. That the plant produces many substances other than food products which may be of value to the manifold needs of industry is, surprisingly enough, only recently coming to realization, and quite naturally promises to effect some basic changes in the relation of agriculture to industry.

In order that the plant may carry on its photosynthetic manufacture it requires an adequate supply of raw materials and a source of energy to propel the process. These raw materials are carbon dioxide and water, which are derived from air and soil; the energy is the light from the sun. One of the remarkable features of the process is that the plant can utilize raw material which may be considered of such very low grade. In the case of land plants, which at least economically are of the greatest significance, they are dependent upon the supply of carbon dioxide in the atmosphere. However, only 0.03 percent of the air is carbon dioxide. On this apparently thin thread hangs all the life of our planet. Under conditions most usually found in nature, if the carbon dioxide content of the atmosphere were to be reduced, the rate of photosynthesis would be correspondingly diminished; and, on the other hand, if, for example, the carbon dioxide content were doubled, the rate of photosynthesis would be increased approximately twofold.

Yet this supply of raw material upon which we must depend for our life's existence is not as frail a thread as would at first appear. The gigantic proportions of the earth's atmosphere constitute a reservoir of carbon dioxide, the magnitude of which is difficult to grasp. The resources of the air alone amount to about 2,200,000,000,000 (2.2×10^{12}) tons; and the amount contained in the sea has been estimated as fifteen to twenty-five times this amount. Some conception of the magnitude of these figures can be gained from a consideration of the fact that, in spite of prodigious quantities of carbon dioxide which have been poured into the atmosphere during the past fifty years through the industrial use of coal and petroleum, no evidence of any increase in the percentage of carbon dioxide in the air has been observed.

The driving force of photosynthesis is the energy, in the form of visible light, from the sun. The computations in the accompanying table may aid in an orientation of the magnitudes of energy which are involved in this process.[‡]

[‡] H. Schroeder, "Quantitatives über die Verwendung der solaren Energie auf Erden," *Die Naturwissenschaften*, 1909, p. 976.

DISTRIBUTION OF TERRESTRIAL ENERGY

Sources	Million calories per year	Calories per year
Solar radiation received at the outside of the earth's atmosphere	1,340,000,000,000,000	1.34×10^{21}
Energy used in the evaporation of water	340,000,000,000,000	0.34×10^{21}
Energy used in photosynthesis, on land, the benthos of the sea, but exclusive of plankton ...	162,000,000,000	0.162×10^{18}
Energy of all flowing water on the earth	50,000,000,000	0.050×10^{18}
Energy of the world's coal production	6,600,000,000	6.6×10^{15}
Utilizable water power ...	2,800,000,000	2.8×10^{15}
Utilized water power	80,000,000	0.08×10^{15}
Total work capacity of the human race	70,000,000	0.07×10^{15}

Figures of this nature have value only in giving some indication of the total activity which is involved, of its relative significance, and possibly of the magnitude of our terrestrial system. Just as the size of a national debt gives little information as to the contribution which every citizen must make toward its liquidation, these figures give little indication of what is going on in the individual manufactories of the plant. And it is important to bear in mind that while the total output of this process is enormous, it is all done by piecework as it were. Each plant represents an individual, itself dependent upon a complex of environmental forces, each leaf of the plant a semi-independent factory of the system, and within each factory millions of units in which the process is actually going on. These units are microscopic organs, the chloroplasts, usually many in each single cell, so that an ordinary sunflower leaf, for example, contains about ten billion of these individual units in which the photosynthetic process occurs. So far as the photosynthetic process is concerned, the rest of the plant, the roots, the stem, and the pores, are merely ancillary organs, which supply the raw materials and carry off the manufactured product. The center of interest for the photosynthetic process is the chloroplast. The chloroplasts are as important to an understanding of this process of photosynthesis as the chromosomes are for heredity. . . .

Of the more obvious components of the chloroplasts we have at least a fair knowledge. The most striking of these is the green coloring matter, chlorophyll. Actually composed of two pigments, the fundamental chemical structure of chlorophyll is now fairly well understood. The more intimate study of the optical and chemical properties of this remarkable substance is revealing a most extraordinary adaptation to the role it plays in the diverse environmental conditions under which the plant must operate in nature. The finer aspects of the properties of chlorophyll are still to be worked out, and will require chemical and physical investigation of the highest refinement. This is particularly true because there is now little doubt that the chlorophyll which has been isolated from the plant and studied in the laboratory differs in some very significant respects from that existing in the plant. In this connection it must be stressed that all efforts to reproduce the photosynthetic process outside the living cell have met with failure.

Besides the chlorophyll there are within the chloroplast other, less obvious pigments, of very different chemical structure. These are the yellow or orange carotenoid pigments. Here is an example of having isolated a part of the machine without knowing just what role it plays in the functioning thereof. Although these pigments are present in every chloroplast, the hypothetical schemes which have been formulated for a mechanism of the reaction cannot with certainty ascribe specific roles to these members of the apparatus. Investigation of the carotenoid pigments has revealed that they constitute a group of chemical compounds of much greater complexity and diversity than was at first supposed. Composed essentially of a long chain of carbon atoms, partially saturated with hydrogen atoms and capable of adding and releasing hydrogen, it is conceivable that they may play a role in the hydrogenation mechanism which is essential to the reduction of carbon dioxide to carbohydrates.

The fact that we can at the present not ascribe a definite chemical or physical role to a particular component of the photosynthetic apparatus is in itself no reason for discouragement. The struggle with this problem of the mechanism of the photosynthetic process has been under way sufficiently long to warrant the conclusion that victory is not going to come by means of a few spectacular and brilliantly executed sorties. It is manifestly a problem demanding the patient strategy of a hard seige.

In 1944 the brilliant and articulate anatomist and embryologist Dr. George W. Corner was invited to give the Terry Lectures at Yale University, on the general subject of embryology and what it can tell us of man's place in the scale of nature. These lectures subsequently became the basis of one of the most delightful books ever written on embryology: Ourselves Unborn, *published in 1944 by the Yale University Press.*

"The Generality and the Particularity of Man" consists of selections from the third and final chapter of that book, written while Dr. Corner was Director of the Carnegie Institution's Department of Embryology, a position which he held from 1940 until 1956. A man of broad scholarship and prolific literary output, Dr. Corner was born in Baltimore, Maryland, in 1889. He received his A.B. degree from the Johns Hopkins University in 1909, and the M.D. degree in 1913. He served as medical assistant to the Grenfell Labrador Mission in 1912 and 1913; was assistant in anatomy at Johns Hopkins in 1913–1914; resident house officer of the Johns Hopkins Hospital in 1914–1915, and assistant professor of anatomy at the University of California from 1915 to 1919. In the latter year he returned to Johns Hopkins as associate professor of anatomy, and he continues his association there as professor emeritus of embryology. He was professor of anatomy at the University of Rochester from 1923 until 1940 and curator of the medical library at that university, and was managing editor of the American Journal of Anatomy *from 1939 until 1941.*

His sixteen years as Director of the Carnegie Department of Embryology were marked by rapid advances in knowledge of human development and the physiology of reproduction, greatly aided by the maintenance of a monkey colony which provided living material for study. During this period also, the Department's collection of human embryos begun by Dr. Corner's predecessor, Dr. Franklin P. Mall, was continued, and today comprises the largest such collection in the world.

Following his retirement as Director of the Department, Dr. Corner served as historian and later as visiting professor at the Rockefeller Institute, and in 1960 became executive officer of the American Philosophical Society.

210

George W. Corner

THE GENERALITY AND THE PARTICULARITY OF MAN

From *Ourselves Unborn*. Copyright, 1944, by
Yale University Press. All rights reserved.

I said in mine heart concerning the estate of the sons of men . . . that
they might see that they themselves are beasts.

Thou hast made him a little lower than the angels, and hast crowned
him with glory and honour.

I once had a caller at my laboratory who spent an
afternoon trying to demonstrate that the human race is the
result of a cross, sometime in the dim past, between an angel
and an anthropoid ape. This hypothesis, which bobs up repeatedly in visionary minds, rests upon a well-established and
on the whole valuable trait of human nature, self-esteem. It
has been all too clear, at every stage of history, that man is
a beast, whether or not he also contains a dash of the angelic.
Yet we are not willing to admit it. Every argument that poetry,
religion, the intellect, or mere racial pride can supply is used
to bolster our self-asserted distinctiveness among living creatures. A man will work and pray and if necessary die for ideals
which imply the claim of special rank for his own kind. Even
those men of science who have tried without prejudice to
determine the place of man in organic nature have most of
them felt a certain restraint in classifying him as an animal.

Scientific thought on this subject began with the idea
that man is a highly superior creature who happens to present
certain resemblances to the animals; it has only very slowly

211

proceeded to the view that he is an animal who happens to possess certain special abilities. Aristotle, for example, in his *History of Animals,* placed man by himself in an exclusive class of living things. He divided the animals somewhat vaguely into those which are bloodless (or as we should now say, the invertebrates), those which have blood (the vertebrates), and man. He said that the principal genera having blood are man, the viviparous quadrupeds, oviparous quadrupeds, birds, and fishes. Man's special powers of thought and action, however, seem to separate him too widely from the four-footed animals to permit including him with them in one class. Some animals, he thought, share the properties of both man and the quadrupeds; these are the monkeys and baboons.

The view that man occupies a special position in nature fits in with an idea about the relationship of living things which was implicit in Aristotle's writings and which influenced the thought of zoologists until the 19th century. Indeed, it still underlies a good deal of casual thinking about the relative place of man and the animals. This is the concept that all animals and plants, in fact all natural objects, inorganic as well as organic, are arranged in a single series or scale of ascending rank, with man at the top. Such a view as this about the human species was well suited to the philosophy of the Middle Ages and the Renaissance, when the supreme value which Christianity set upon man's immortal soul strengthened the tendency to regard him as a very distinctive kind of being, occupying the chief place in earth's hierarchy. "Man, the most excellent and noble creature of the world," says Robert Burton in the *Anatomy of Melancholy,* "the principal and mighty work of God, wonder of nature . . . *audacis naturae miraculum,* the marvel of marvels . . . the abridgement and epitome of the world . . . sovereign lord of the earth, sole commander and governor of all creatures in it; . . . far surpassing all the rest, not in body only, but in soul."

Such a rhapsody, and indeed the whole proud assumption that man is different to an immeasurable degree from other animals, might have been less confident, and the battle between science and orthodoxy over the question might have begun far earlier than the 19th century, if European man had known anything about his higher simian relatives. Had the chimpanzee or gorilla been available for anatomical comparison, mankind would not have seemed so far in advance of his fellow creatures. Until the 17th century there were only vague stories about man-like animals in distant Africa and Asia. The first of the great apes known to have been brought alive to Europe was a chimpanzee described in 1641 by Nicholas Tulp, the Dutch anatomist who is the leading figure in Rembrandt's painting "The Anatomy Lesson." The earliest recorded dissection of a great ape was reported in a remarkably accurate and sensible work by Edward Tyson, published by the Royal Society of London in 1699. Tyson made his opinion perfectly clear that the body of this animal — he too was studying, like Tulp, a young chimpanzee — resembled that of man far more closely, as he said, "than any of the ape kind, or any other animal in the world, that I know of." Evidently, then, man has closer relatives in the anatomical scale than had been realized before; a fact which thenceforth had to be taken account of by those who undertook the classification of animals. Théophile Bonnet, for example, an industrious French naturalist of the 18th century, published a detailed tabulation of the Scale of Beings, beginning with Fire, Air, and Water, through the ascending ranks of sulphurs, metals, stones, corals, molds, lichens, plants, the various invertebrate and vertebrate animals, upward to the quadrupeds, monkeys, chimpanzee, and man. All that such a list requires to make it reach from the lowest to the Highest, is to add the celestial hierarchy of angels, archangels, powers, and principalities. Man may thus be considered either the highest of beasts or the lowest of the heavenly hosts, an arrangement which permits him to be dealt with by both biology and eschatology without conflict.

The idea of the Scale of Beings broke down, however,

with the wider exploration of nature. As the naturalists built up an even larger list of species, it became absurd to try to rank them in serial order. Who shall say whether the fox is higher than the wolf, or a sparrow than a robin? After the appearance in 1735 of Linné's *System of Nature* the science of classification was revolutionized. The scalar arrangement gave way to the newer and more manageable method of classification by classes, orders, and families. Linné himself never saw one of the great apes, but Tyson's monograph had made it necessary for him to find a place for the chimpanzee which should not be very far from man, on the one hand, or the monkeys on the other. He therefore defined seven orders of mammals, one of which he called* the *Primates,* comprising man, the apes and monkeys, the lemurs, and the bats. His reasons for grouping the bats so closely with the others need not concern us here; they were not valid, and later zoologists placed these creatures in a special order, the Chiroptera. Whether or not the lemurs are enough like the apes and monkeys to be placed in the same order with them is still actively debated. It is a question on which the embryologists have something to say and we shall return to it later.

Not all the competent zoologists after Linné were willing to follow his bold consideration of man as a member of the same order of mammals as the apes and monkeys. Johann Friedrich Blumenbach, the founder of physical anthropology, in his classification of 1779 placed man in a separate order which in a later work he called *Bimana,* two-handed, in contrast to the other apes and monkeys, the *Quadrumana,* four-handed. Thomas Pennant, a good English naturalist, in 1781 wrote, "I reject Linné's first division, which he calls Primates or Chiefs of Creation; because my vanity will not suffer me to rank mankind with Apes, Monkeys, Maucaucos and Bats, the companions Linnaeus has allotted us even in his last system." Baron Cuvier's great *Règne Animal* of 1817 still maintained

* In the 10th and later editions, 1758 ff.

Blumenbach's special Order of Bimana, occupied solely by man in all his pride.

All this may seem a merely technical question, but the British-American public of mid-Victorian days did not think so. The outburst of an open controversy, about the year 1860, on the zoological position of man was surely one of the great turning-points of human thought. . . . By the time Huxley had finished his debates with Bishop Wilberforce and Professor Owen the attitude of every educated person respecting our place in nature was permanently affected. No philosophy of life, no system of ethics could any longer neglect the animal nature of man. Three circumstances brought the question to a head. One was the progress of geographic exploration, through which knowledge of the great apes came flooding into the zoological gardens, museums, and dissecting rooms of Europe and America. Travelers in eastern Asia and Africa sent home specimens of the chimpanzee, gorilla, and orang-utan. Richard Owen, the great comparative anatomist at the Royal College of Surgeons of London and afterward at the British Museum, was thus enabled to give a comprehensive description of the skeletons of the chimpanzee and orang-utan in 1835. An American medical missionary, Thomas S. Savage, collected skeletons of the gorilla in 1847 from which Jeffries Wyman of Harvard worked out the osteology of that species. The gibbons, smallest of the man-like apes, and known in the 18th century only from one specimen described by Buffon, were also made the subjects of anatomical investigation. Thus it became well understood that there are five genera of anthropoid apes, forming a group clearly distinguishable from the monkeys, and in some characteristics more like man.

Another factor was the discovery of remains of fossil men, especially the sensational find at Neanderthal in 1857, which proved the existence in prehistoric times of a now-extinct species of man of coarser build and lower brow than modern man. Thus it appeared that the gap between man and the monkeys was being filled from both ends.

The third reason for the sudden importance of the question of man's zoological position was the announcement by Darwin and Wallace in 1858 of the theory of evolution by natural selection. Mere anatomical facts, or even the discovery of fossil man, might have attracted little immediate attention outside the laboratories and museums; but the publication of Darwin's *Origin of Species* made a public furore. Its implication that the human species is descended from an ape-like ancestor was attacked on grounds both of religion and of science. Some conservatively minded anatomists came to the front with observations that seemed to prove wide differences between the anatomy of man and that of the great apes. Richard Owen in particular had long since described supposedly fundamental divergencies between the skulls and brains of man and of the great apes, thus apparently confirming the classification of his teacher Cuvier, by which man was set apart as the only species and genus of the Order Bimana. This anatomical observation, now brought forward in opposition to the Darwinian hypothesis, was answered fully and devastatingly by Huxley, together with other arguments in *Man's Place in Nature* (1863), one of the masterpieces of scientific exposition in English, in which he showed that in many anatomical features the differences between man and the great apes are actually less than those existing between these apes and the monkeys.

In spite of all that he did to emphasize the distinctive features of the great apes† as contrasted with the monkeys, Huxley did not set them up in his classification as a separate suborder or even as a separate family. In his *Anatomy of Vertebrated Animals* (1872) he ranked the primates as shown below. It will be seen that he gave man much the same relatively distinctive position as had Linné.

---

† For the sake of clarity, the term *anthropoid apes* is used here to include the gibbon, siamang, orang-utan, gorilla, and chimpanzee; *great apes* means only the last three.

Order	Suborder	Family	Subfamily

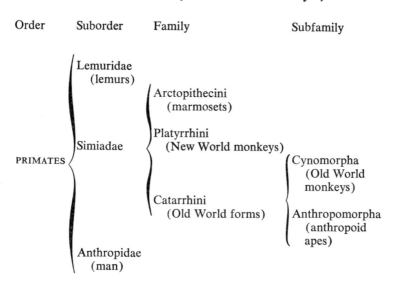

Lemuridae
(lemurs)

Arctopithecini
(marmosets)

Platyrrhini
(New World monkeys)

PRIMATES

Simiadae

Cynomorpha
(Old World
monkeys)

Catarrhini
(Old World forms)

Anthropomorpha
(anthropoid
apes)

Anthropidae
(man)

It remained for Sir William Henry Flower to grant (1883) full recognition to the anthropoid apes as a group so similar to one another, and so different from the monkeys, that they must be placed on the same level of classification with the Old World monkeys. Thus as increasing information has been obtained about the primates, the anatomists have gradually elevated the anthropoid apes to a higher level of classification, at the same time ranking the human species more and more intimately with the other animals of similar type. In Aristotle's time man was set apart from all the other vertebrates. Linné included him with them and grouped him with the primates, but in a separate suborder as we should now say; Huxley set forth the claim of the great apes as his near relatives; and 20th-century biology, rich in knowledge of primate species, considers living and fossil man as only one family among four which form the tribe of Old World apes and monkeys, thus:

> Infraorder Catarrhini:
> Family Cercopithecidae (Old World monkeys)

> Family Hylobatidae (gibbon, siamang)
> Family Pongidae (orang-utan, gorilla, chimpanzee)
> Family Hominidae (man, living and fossil)

If this slow but steady demotion of man (as judged by the biologists) were to continue, the next step would obviously be to throw over the last quantum of human pride and classify our species merely as one of the subfamilies of the catarrhine apes; in other words to say that we are simply one particular kind of ape or monkey. There are at this moment biologists who would do just that, but the differences and resemblances upon which such a decision must be based are so diverse that it becomes a matter of private judgment, hardly subject as yet to general agreement, whether to place the human species with the great apes or with the monkeys.

Punch's famous query, made at the height of the Huxley-Owen controversy in 1861, quite literally expresses (excepting the first line) the question now awaiting the answer of science:

> Am I satyr or man?
> Pray tell me who can,
> And settle my place in the scale;
> A man in ape's shape,
> An anthropoid ape,
> Or a monkey deprived of a tail?

The majority of investigators, however, still think that man is sufficiently distinctive to constitute a family by himself. After all, if he is an ape he is the only ape that is debating what kind of ape he is.

What evidence has embryology to bring to bear upon this great problem? . . .

The embryos of various species of mammals can be distinguished from one another by the expert observer, all the way back to the stages of the one-celled ovum, and there is an endless variety in the manner of implantation, in the arrangement of the embryonic membranes, and in the structure of the placenta. Is it too much to hope that by careful comparison and codification of these differences a flood of light may be thrown upon the problem of human descent?

In the first flush of enthusiasm over their discoveries of the early 19th century, embryologists seemed indeed ready to solve all the tangled questions of evolution, if only they could apply their methods to the whole animal kingdom. Von Baer, the man of genius who in 1827 discovered the eggs of mammals, made it his business to collect the embryos of all sorts of vertebrates and invertebrates. He perceived that an early embryo shows the general characteristics of the big group to which it belongs, before it acquires the special characters of its genus or species. A good example of this is seen in the fore-limb buds of birds and mammals, which are very much alike when they first grow out of the trunk; it is only later that they are distinguishable as wings or legs respectively. To cite a more general example, all the many-celled animals begin as a single cell, and practically all of them, vertebrates and invertebrates alike, go through a morula stage and then a blastocyst stage or something much like it, before they develop the particular characteristics of their own kind. Thus animals of higher and lower forms resemble each other much more when they are embryos than when they are adults. This is another way of saying that they all start out more or less alike and diverge as they grow. It follows that the adults of lower, less differentiated forms which do not get very far from the embryonic, generalized type will to some degree resemble the embryos of higher, more divergent forms.

In the hands of Haeckel, whose facile writings had a great influence on popular thought about evolution in the latter half of the last century, these principles of von Baer were built into the so-called "biogenetic law" or theory of recapitulation, which declared that higher animals, in their embryonic stages, undergo the same series of transformations through which their ancestors passed during the evolution of the species. Mammalian embryos, for example, were thought to exhibit a fish stage and an amphibian stage, and Haeckel at one time went so far as to draw up a table of the progenitors of man which consisted, practically speaking, of a list of adult forms of now-existent types, all the way up from the invertebrates, i.e., segmented worm, cyclostome (lamprey), shark, bony fish, amphibian, marsupial, lemur, monkey, and ape. If such an idea were true in detail, the embryologist could solve all the relationships of the animal kingdom by studying the development of each species to see what kinds of animals their embryos successively resembled, but even those who cherished the notion most warmly had to agree that in the development of an embryo many stages of evolutionary history are omitted or greatly condensed. The human embryo is never a monkey or a reptile, nor is it ever like an *adult* fish or amphibian. It does, however, show remarkable resemblances to *embryonic* fishes and amphibians. Microscopic study of the internal organs, as depicted in any textbook of vertebrate embryology, gives convincing evidence of the fact. Yet such is the persistence of theoretical ideas that to this day some of the visitors to an embryological laboratory are actually disappointed to find that human embryos are not sufficiently like adult fish to exhibit fully formed gill slits, to say nothing of fins and scales.

Let us see then what the embryologist can contribute toward understanding the position of man among the animals. Differences in the placenta of various animals are highly significant. That organ in its gross form may be either diffuse, cotyledonary, zonary, discoidal, doubly discoidal, or some other intermediate or variant shape. In its general structure it may be of the diffusely apposed type, or labyrinthine, or villous. In detailed microscopical structure, the placenta exhibits four general types according to the number of cell layers which remain interposed between the blood streams of mother and fetus. The placenta has therefore long been subjected to careful study with the aim of comparing one species with another. An expert in this field, with slides of all the known placentas before him, would be able to separate those of Tarsius and the pithecoid primates (monkey-ape-man) from those of all other animals without difficulty, and could then go on with ease and confidence to identify those of the great apes and man because of their striking interresemblance. He might hesitate however before committing himself to a decision as to which came from the gorilla and which from the human species. With respect to the structure of the mature placenta, man is thus definitely closer to the great apes than to any other animals.

But the organization of the placenta is an outcome of events which take place in early embryonic life. Even in the first two weeks there are great differences in the method of attachment of the embryo to the mother and in the arrangement of the embryonic membranes. The detailed variations of these early stages of development are so pronounced that the total picture of implantation and the arrangement of the membranes, in any one kind of animal, becomes an important item in the list of characteristics available for taxonomy and evolutionary comparison.

The relevant information is fairly well in hand as regards the human embryo. Fortunately the early stages of the rhesus monkey also are quite fully known, thanks to the intensive work of Streeter, Hartman, and Heuser of the Department of Embryology of the Carnegie Institution, in Baltimore, who have made the development of this monkey better known than that of almost any other mammal. The human embryo and that of the monkey are much alike during the second week, in those details which are significant for comparative study. Both show precocious development of the trophoblast (considerably more pronounced in the human), formation of the amnion by splitting, and early development of the extraembryonic mesoblast. They differ significantly, however, in the type of primary implantation, which is superficial in the monkey, interstitial in the human. With this understanding of the earliest stages of the human species and one of the most typical of the Old World monkeys, we are prepared to turn to the great apes for a comparison which should tell us much.

Alas, the total material for the study of the early development of the great apes consists of two embryos of the chimpanzee, 10½ and 12 days old respectively, at the Carnegie embryological laboratory. The older of the two is very abnormal and thus almost everything that can be said on this great subject depends upon one embryo. The two specimens were obtained in 1938–39 by coöperation of the Yale (now Yerkes) Laboratories of Primate Biology at Orange Park, Florida. Operations by Carl G. Hartman were done on dates worked out from the menstrual cycles on the basis of previous studies on the reproduction of the chimpanzee by R. M. Yerkes and J. H. Elder, and the specimens were prepared and intensively studied by Chester H. Heuser (1940). Their importance with respect to the relation between chimpanzee and man was first pointed out by Dr. Hartman in 1939.

The findings in the case of the 10½-day specimen ("Yerkes A," Carnegie Collection C.620) can be stated in a few words. The embryo is practically identical with human embryos of similar age in every significant detail. The implantation is interstitial and is characterized by very early thickening of the trophoblast. There are certain differences from the human in slight details which may reflect the difference in species, but it is doubtful whether even a trained embryologist, seeing it for the first time, could detect the fact that it is not human.

In short, at the end of 10 or 11 days the human species and one of the great apes (the only one known at that age), are anatomically alike and are different in significant details from the rhesus monkey of a comparable stage and development, although in most respects the embryos of the three species are very similar and must be placed together in a group differing from embryonic mammals of other orders.

It may be added that enough is known about the embryos of the gibbon and siamang (Hylobates, Symphalangus), beginning at a stage many days later than the Yerkes chimpanzee embryo, to make it certain that in these apes the implantation is interstitial, like that of man. The same deduction can be made from the late placentas of the gorilla and orangutan.

Before attempting to evaluate these remarkable facts, it will be worth while to examine very briefly what is known of the early embryology of other primates. As the reader will realize, there are exceptional difficulties in the way of collecting monkey and ape embryos. To get together a useful series of successive stages of any species by collection in the field requires either special expeditions or the cooperation of scientifically minded residents of the tropics. Pregnant animals, shot or trapped in the jungle, are of course not subject to accurate determination of the stage of pregnancy. To breed monkeys in captivity expressly to get the embryos is an expensive business. Only a few species are hardy in captivity. We know almost nothing about the breeding habits of most of the monkeys. Thus far, only the Carnegie series of rhesus embryos and the two Yerkes chimpanzee embryos have been obtained in that way. Considering these difficulties it is remarkable how much has been discovered by an international group of men, little known to the general public, who have devoted themselves, in the field and in the laboratory, to this bold and far-reaching quest. Emil Selenka (1842–1902) of Erlangen, Germany, opened the modern phase of the subject by his studies of material obtained largely by personal travel in South America and Asia. A. A. W. Hubrecht (1853–1915) made extensive use of his personal connections in the Dutch East Indies, and of his private means, to build up a magnificent collection of primate and insectivore embryos preserved, since his death, in a special institute in his late residence at Utrecht, Holland. Hans Blunt-schli of Bern, Switzerland, made valuable collections of New World monkeys during travels in Amazonia. I have already had occasion to mention James Peter Hill, of London, who made himself a master of marsupial embryology during a long career in Australia and contributed greatly to knowledge of the primates; George B. Wislocki, who advanced our under-standing of placentation by studies made at Johns Hopkins and later at Harvard Medical School, Boston, based in part on his own collections in Central America; and George L. Streeter, Carl G. Hartman, and Chester H. Heuser, whose magnificiently successful investigation of the embryology of the rhesus mon-key was carried out in our Carnegie Laboratory in Baltimore from 1923 to 1939. . . .

Summarizing the work of these and other investigators, we must begin by stating that the lemurs differ from all the other primates in their method of attachment of the embryo to the mother. Moreover, enough is known of the early embryol-ogy of various lemurs to show that the amnion is formed by folding and that the allantois attains a fair size and takes part in forming the placenta. In these features also the lemurs differ from the other primates.

Tarsius is next to be considered, that eerie wraith of the East Indian forests, who bears in his tiny body such vaguely disquieting signs of relationship to man that the natives of Sumatra and Borneo stand in awe of him as if he were a ghost, and so for different reasons do occidental men of science, to whom he seems no ghost of a dead past but a portent of the rise of man. The first tarsiers ever brought away alive from their native lands were obtained through the efforts of John F. Fulton of Yale University, who has written an interesting account of his trip to the Philippine Islands in 1938. A. A. W. Hubrecht, who worked out the embryology of Tarsius, never saw a living specimen, although he collected more than 450 embryos through contacts he made in the Dutch islands. From his work we know that the embryological development of this primitive primate shows several features that are also characteristic of the higher primates. The blastocyst, for example, very early acquires a direct attachment to the uterine lining. The trophoblast spreads out upon and in the superficial layer of the endometrium somewhat as does that of the rhesus monkey. The placenta is massive, discoidal, and of the deciduate hemochorial type. The extraembryonic mesoblast develops precociously, giving rise to a body stalk much as in man; and the allantois is never more than a small duct in the body stalk. The yolk sac is small. Unlike the higher primates, however, the amnion forms late by folding of the chorion over the body of the embryo.

In the New World monkeys, which are known only from relatively scanty collections, the amnion forms by splitting; the yolk sac apparently develops in some such way as in man and the rhesus monkey; and the extraembryonic mesoblast is formed even more precociously than in Tarsius. The allantois is rudimentary. The placenta, which in most species forms in two masses at opposite poles of the chorionic sac, is massive and hemochorial.

The researches of Streeter and Heuser on the rhesus macaque, typical of the Old World monkeys, have shown that in all these details we are now discussing, the monkey resembles man, the only striking difference being that the monkey's blastocyst does not burrow at an early date below the surface of the endometrium, but instead the trophoblast at first invades the surface layer and then spreads out to form the placenta. We have already seen that the great apes, as far as we can be certain from a few placentas of the orang-utan and one of the gorilla and one early embryo of the chimpanzee, are so similar to man in early development and placentation as to be almost indistinguishable.

Summing up all the available information from the embryo and the placenta, it may fairly be said that if the embryologist were called upon to state his views about the classification of the primates without any knowledge of the anatomy of the adult, he would assemble them in three groups, one containing man and the great apes, one containing the Old World and the New World monkeys, and one containing Tarsius. A commonly accepted conjecture at present is that the lemurs are only collateral relatives of the pithecoid (ape-monkey) line. . . .

It is, of course, not safe to choose any one characteristic, or any limited group of features, upon which to base a theory of relationship among animals. Fundamental and important as the type of implantation and placentation admittedly is, it cannot outweigh all the other evidences from bones and muscles, brains, teeth, and whatever else the comparative anatomists may be able to cite. The findings of the embryologist must be used in combination with all other available facts.

The embryology of our own species, as far as we can compare it with our scanty knowledge of other primates, associates man even more intimately with the great apes than does the evidence from comparative anatomy of the adult body. The embryologists, left to themselves, could hardly fail at present to support the most orthodox Darwinian-Huxleyan doctrine of the descent of man from an ape-like ancestor, in close relation with the gorilla and the chimpanzee. Knowing, however, the limitations of an approach through any one branch of science in summing up the evidence in the case, all they will insist upon is that the human embryo is that of a primate closely similar to the apes and monkeys. As far as we embryologists are concerned, the comparative anatomists, when they refine their diagrams, may choose to split off the human line either close to the point where the monkeys separated themselves, or close to the stem of the great apes. When we can get early embryos of the gorilla, the orang-utan, and the gibbon, and more stages of the chimpanzee, we may be able to add something more precise.

Many comparative anatomists at the present day would agree in considering the insectivores the most primitive of living mammals and in supposing that the earliest placental mammals must have been animals of that general sort.

Since both comparative anatomy of the adult and embryology agree in seeing many resemblances between the insectivores and the line of Tarsius, monkeys, apes, and man, it follows that our human body in its development and perhaps in its adult structure may have preserved somewhat more of the general primitive mammalian type than many of those animals we are pleased to consider our inferiors. Let us pursue this thought in the concluding part of our study of the natural history of the human embryo.

The human body is indeed surprisingly generalized; that is to say, man is in most anatomical features a typical and unpeculiar vertebrate. As the Vertebrata radiated into their various classes, the Class of Mammalia into its orders, and the Order of Primates into families, genera, and species, the ancestors of the Hominidae were among those which (relatively speaking) held to the middle way, not developing extensive specialization. This is crudely apparent when one considers that there is scarcely any purely physical activity in which man is not excelled by some other animal — neither running, swimming, diving, flying, climbing, nor in withstanding heat and cold, nor in the use of any of the five senses. He can, it is true, stand and walk on his hind legs more steadily than any other. There is no space here to work out the full technical argument for the generality of the human body. F. Wood Jones in a forceful passage in *Man's Place Among the Mammals* (1929) points out that the Primate Order is one without distinction, but with essentially generalized mammalian features. It would be going much too far to say that man, a member of this order, is himself without some structural distinction, but it is generally agreed by comparative anatomists that he is not a creature of outstandingly special traits.

It is the brain to which we must look if we would find an organic basis for the self-rated superiority of man, but even this organ is built upon the common plan of the vertebrate brain, and as far as gross anatomy goes is so closely similar to that of the great apes that the differences cannot be qualitatively defined. Sir Grafton Elliot Smith's remark has often been quoted, that "No structure found in the brain of an ape is lacking in the human brain; and on the other hand, the human brain reveals no formation of any sort that is not present in the brain of the gorilla or chimpanzee." The large size of the human brain, however, and the altered quantitative relation by which the cerebral hemispheres are relatively larger in proportion to the rest of the brain than in other primates, comprise of course a kind of specialization as significant as if the brain had developed some new lobe unknown in the rest of the animal kingdom. Thus the relatively unspecialized human body is guided by a brain which is functionally superior to that of other animals. The creature that owns this combination cannot run or swim as well as many of his fellow creatures, nor can he fly at all without mechanical assistance; but his specialized brain can guide his generalized and therefore generally adaptable body to build fast-moving cars and ships, and to fly, with a machine, longer and higher than any bird.

After so long a preamble I come at last to recapitulate what the embryologist has to say about this question of the generalization versus the specialization of man. In the past, human development has been regarded as a very special affair. The rarity and preciousness of early human embryos seem to confer distinction upon them, but as knowledge of a wide range of animals has been obtained early human development has come to be more and more understandable in terms of the general mammalian plan.

It is not my purpose here to continue the theme into the later stages of human development. In fetal life as in the embryo there are numerous details of bodily structure and proportion which strongly support the idea of a common origin for all the primates, and also the conclusion that the human body is in many respects more primitive than that of various other primates.

In summary it may fairly be said that the early embryology of the primates brings forward no observations which stand in necessary disagreement with the concept of man as a relatively generalized animal, which has been arrived at by the study of comparative anatomy and of fetal structure.

In the foregoing pages I have outlined the evidence for two very weighty conclusions about the physical nature of man. The first of these is that he is an animal, a member of the Class of Mammalia, Order of Primates, and closely related to the apes and monkeys. The second is that the human body is not notably endowed with specialized anatomical features of a kind that would fit us to perform limited activities supremely well, but on the contrary is built rather closely to the general mammalian pattern, and therefore can perform varied tasks under the guidance of a superior brain. As the reader must have noticed, what I have said about the contribution of embryology to this subject, which after all is the real theme of the discourse, has largely consisted in the removal of embryological objections to these two conclusions. The difficulties which delayed the investigation of early human development, together with a certain prideful if often unconscious expectation that human structure must always exhibit superiority to other animals, even in its earliest gestation, have given a false impression that the human embryo is exceptional.

Let us pause a moment to reflect that generalized organisms possess a degree of liberty in action which is denied to more specialized creatures. The evidence from embryology points to a kind of foreordination, not exactly that of the theologians but none the less operative to determine the fate of countless human beings. We are led, on the other hand, by the evidence from comparative anatomy to ponder upon the freedom of the will, or at least freedom of action, which we have because our bodies are versatile, untrammeled by specialization for extreme but particular skill, and capable of any task the mind may imagine. The way of an eagle is to fly, the way of a fish is to swim, but a man's way is to kill or to cultivate, to grovel or to stand erect, to dissipate or to build. Because he has a generalized body he can eat any diet, live in jungle, prairie, or upland, swim across rivers or scale mountains, build altars to many gods. If anyone insists that all this freedom inheres in the mind alone, let him name a valiant deed or a crime that does not depend also upon the anatomical versatility of our bodies, or cite any great life that could have been lived by a man with bat's wings or the alimentary canal of a sheep. The scope of the human mind, the freedom of human decision, are bound up inextricably with the generalization of the body.

The animal nature of man is an old story, as old or older than the Scripture with which I have adorned the title of this essay; but it must be retold in each age and restudied in the light of all that science can learn, for the whole structure of our laws, our philosophy, and our religion depends upon the way in which we look at human nature. We shall always have among us, of course, those who decry the body and depreciate its claims in the interest of the soul; among them, be it said, many noble persons, gaunt saints and shining heroes who have taught their fellow men how to glorify the flesh by following the spirit. And yet such deeds as theirs, and every act of love and sacrifice, even man's dreams of beauty and his prayers to heaven, must be realized with the flesh, bones, and organs of an animal that is blood-brother to the worm and the ape.

Who should know this better than the physician-scientist whose privilege it is to observe the frail minuscular germ of our race in its earliest days, and who sees constantly in his work what I have tried to tell in these pages: that we begin our lives in continuance of a long past and in progression toward an unseen goal; that life is precarious from the first day to the last, under the sway of events we can neither comprehend nor calculate; and that we bear through all our days the marks of intimate kinship with the animal world, tempered by powers of the mind that bestow dignity and honor upon the life of the body.

IV
THE ROLE OF GREAT INSTRUMENTS

In 1927 the editor of Harper's Magazine *asked Dr. George Ellery Hale to write an article on astronomy. Hale agreed, and took the opportunity to write "The Possibilities of Large Telescopes," in the hope that he might stimulate others with his own enthusiasm. While the article was still in proof he sent a copy of it to Wickliffe Rose of the Rockefeller Foundation. Much to his surprise, Rose answered immediately: "It is a matter that interests us. We should be very glad to discuss it with you." It was thus that the 200-inch Hale telescope on Palomar Mountain, which began operation in 1948, came into being.*

This was not the first time that Hale had managed to turn his visions into realities. Born in 1868 in Chicago, and educated at M.I.T., he was the organizer of an observatory of his own, and its director (the University of Chicago's Yerkes Observatory). Under his direction, the Yerkes 40-inch refracting telescope was built. In 1904, just two years after the founding of the Carnegie Institution of Washington, he organized the Institution's Mount Wilson Observatory and remained its Director until 1923. Establishing the Observatory with two small telescopes borrowed from the Yerkes Observatory, Hale was accustomed to make his way up the mountain on foot with a mule to carry his equipment.

Through his efforts, the facilities of the Observatory improved rapidly. In 1904 he received a grant of $300,000 from the Institution to supplement money given him by his father, and began construction of the 60-inch telescope. Just six years later he received gifts from John D. Hooker, a Los Angeles businessman, and Andrew Carnegie, and built the 100-inch telescope.

Dr. Hale died in 1938. During the course of his seventy years he brought four major telescopes into existence, each the world's largest and most powerful at the time it was built. He brought to the science of astronomy almost unrivalled energy and vision, and his contributions to our understanding of the universe marked him as one of the world's outstanding leaders in astronomical research.

234

George Ellery Hale

THE POSSIBILITIES OF LARGE TELESCOPES

Copyright 1928 by Harper's Magazine, Inc.; copyright renewed 1956 by Harper's Magazine, Inc. Reprinted from April 1928 issue of *Harper's Magazine* by permission of the author's estate.

Like buried treasures, the outposts of the universe have beckoned to the adventurous from immemorial times. Princes and potentates, political or industrial, equally with men of science, have felt the lure of the uncharted seas of space, and through their provision of instrumental means the sphere of exploration has rapidly widened. If the cost of gathering celestial treasure exceeds that of searching for the buried chests of a Morgan or a Flint, the expectation of rich return is surely greater and the route not less attractive. Long before the advent of the telescope, pharaohs and sultans, princes and caliphs built larger and larger observatories, one of them said to be comparable in height with the vaults of Santa Sophia. In later times kings of Spain and of France, of Denmark and of England took their turn, and more recently the initiative seems to have passed chiefly to American leaders of industry. Each expedition into remoter space has made new discoveries and brought back permanent additions to our knowledge of the heavens. The latest explorers have worked beyond the boundaries of the Milky Way in the realm of spiral "island universes," the first of which lies a million light-years from the earth while the farthest is immeasurably remote. As yet we can barely discern a few of the countless suns in the nearest of these spiral systems

235

and begin to trace their resemblance with the stars in the coils of the Milky Way. While much progress has been made, the greatest possibilities still lie in the future.

I have had more than one chance to appreciate the enthusiasm of the layman for celestial exploration. Learning in August, 1892, that two discs of optical glass, large enough for a forty-inch telescope, were obtainable through Alvan Clark, I informed President Harper of the University of Chicago, and we jointly presented the opportunity to Mr. Charles T. Yerkes. He said he had dreamed since boyhood of the possibility of surpassing all existing telescopes, and at once authorized us to telegraph Clark to come and sign a contract for the lens. Later he provided for the telescope mounting and ultimately for the building of the Yerkes Observatory at Lake Geneva, Wisconsin.

In 1906 Mr. John D. Hooker of Los Angeles, a business man interested in astronomy, agreed to meet the cost of making the optical parts for an 84-inch reflecting telescope in the shops of the Mount Wilson Observatory in Pasadena, where a 60-inch mirror had recently been figured by Ritchey. Before the glass could be ordered he increased his gift to provide for a still larger mirror. Half a million dollars was still needed for the mounting and observatory building, and Mr. Carnegie, who was greatly taken with the project during his visit to the Observatory in 1910, wanted the Carnegie Institution of Washington to supply it. The entire income of the Institution was required, however, to provide for the annual expenses of its ten departments of research, of which the Observatory is one. Nearly a year later I was on my way to Egypt. At Ventimiglia, on the Italian frontier, I bought a local newspaper, in which an American cable had caught my eye. Mr. Andrew Carnegie, by a gift of ten million dollars, had doubled the endowment of the Carnegie Institution of Washington. A paragraph in his letter to the Trustees especially appealed to me: "I hope the work at Mount Wilson will be vigorously pushed, because I am so anxious to hear the expected results from it. I should like to be satisfied before I depart, that we are going to repay to the old land some part of the debt we owe them by revealing more clearly than ever to them the new heavens."

I hope that the 100-inch Hooker telescope, thus named at Mr. Carnegie's special request, has justified his expectations. Its results, described in part in *The New Heavens, The Depths of the Universe,* and *Beyond the Milky Way* have certainly surpassed our own forecasts. They have given us new means of determining stellar distances, a greatly clarified conception of the structure and scale of the Galaxy, the first measures of the diameter of stars, new light on the constitution of matter, new support for the Einstein theory, and scores of other advances. They have also made possible new researches beyond the boundaries of the Milky Way in the region of the spiral nebulae. Moreover, they have convinced us that a much larger telescope could be built and effectively used to extend the range of exploration farther into space. Lick, Yerkes, Hooker, and Carnegie have passed on, but the opportunity remains for some other donor to advance knowledge and to satisfy his own curiosity regarding the nature of the universe and the problems of its unexplored depths.

El Karakat, an Arabian astronomer who built a great observatory at Cairo in the twelfth century, once exclaimed to the Sultan, "How minute are our instruments in comparison with the celestial universe!" In his day the amount of light received from a star was merely that which entered the pupil of the eye, and large instruments were constructed, not with any idea of discovering new celestial objects, but in the hope of increasing the precision of measuring the positions of those already known. Galileo's telescope, which suddenly expanded the known stellar universe at the beginning of the seventeenth century, had a lens about 2¼ inches in diameter, with an area eighty times that of the pupil of the eye. This increase in light-collecting power was sufficient to reveal nearly half a million stars (over the entire heavens), as compared with the few thousands previously within range. The 100-inch mirror of the Hooker telescope, which collects about 160,000 times as much light as the eye, is capable of recording photographically more than a thousand million stars.

While the gain since Galileo's time seems enormous, the possibilities go far beyond. Starlight is falling on every square mile of the earth's surface, and the best we can do at present is to gather up and concentrate the rays that strike an area 100 inches in diameter. From an engineering standpoint our telescopes are small affairs in comparison with modern battleships and bridges. There has been no such increase in size since Lord Rosse's six-foot reflector, completed in 1845, as engineering advances would permit, though advantage has been taken of the possible gain in precision of workmanship. The time thus seems to be ripe for an examination of present opportunities, which must be considered in the light of recent experience.

I have never liked to predict the specific possibilities of large telescopes, but the present circumstances are so different from those of the past that less caution seems necessary. The astronomer's greatest obstacle is the turbulence of the earth's atmosphere, which envelops us like an immense ocean, agitated to its very depths. The crystal-clear nights of frosty winter, when celestial objects seem so bright, are usually the very worst for observation. Watch the excessive twinkling of the stars, and you will appreciate why this is true. In a perfectly quiet and homogeneous atmosphere there would be no twinkling, and star images would remain sharp and distinct even when greatly magnified. Mixed air of varying density means irregular refraction, which causes twinkling to the eye and boiling images, blurred and confused, in the telescope. Under such conditions a great telescope may be useless.

This is why Newton wrote in his *Opticks:*

> If the Theory of making Telescopes could at length be fully brought into practice, yet there would be certain Bounds beyond which Telescopes could not perform. For the Air through which we look upon the Stars, is in a perpetual Tremor; as may be seen by the tremulous Motion of Shadows cast from high Towers, and by the twinkling of the fix'd stars. The only remedy is a most serene and quiet Air, such as may perhaps be found on the tops of the highest Mountains above the grosser Clouds.

Even at the best of sites, in a climate marked by long periods of great tranquillity, unbroken by storms, the atmosphere remains the chief obstacle. For this reason we could not be sure how well the 60-inch and 100-inch reflecting telescopes would work on Mount Wilson until we had rigorously tested them. Large lenses or mirrors, uniting in a single image rays which have traveled through widely separated paths, are more sensitive than small ones to atmospheric tremor. So it has always been a lottery, as we frankly told the donors of the instruments, whether the next increase in size might not fail to bring the advantages we sought.

Fortunately we have found, after several years of constant use, that on all good nights the gain of the 100-inch Hooker telescope over the 60-inch is fully in proportion to its greater aperture. The large mirror receives and concentrates in a sharply defined image nearly three times as much light as the smaller one, with consequent immense advantages. But the question remains whether we could now safely advance to an aperture of 200 inches, or, better still, to 25 feet.

Our affirmative opinion is based not merely upon the performance of the Hooker telescope, but also upon tests of the atmosphere made with apertures up to 20 feet. The Michelson stellar interferometer, with which Pease has succeeded in measuring the diameters of several stars, is attached to the upper end of the tube of the Hooker telescope. When its two outer mirrors are separated as far as possible, they unite in a single image beams of starlight entering in paths 20 feet apart. By comparing these images with those observed when the mirrors are 100 inches or less apart, Pease concludes that an increase of aperture to 20 feet or more would be perfectly safe. For the first time, therefore, we can make such an increase without the uncertainties that have been unavoidable in the past.

Other reasons that combine to assure the success of a larger telescope are the remarkable opportunities for new discoveries revealed by recent astronomical progress and the equally remarkable means of interpreting them afforded by recent advances in physics.

These new possibilities are so numerous that I must confine myself to three general examples, bearing upon the structure of the universe, the evolution of stars, and the constitution of matter. A 200-inch telescope would give us four times as much light as we now receive with the 100-inch, while a 300-inch telescope would give nine times as much. How would this help in dealing with these questions?

The first advantage that strikes one is the immense gain in penetrating power and the means thus afforded of exploring remote space. The spiral structure of nebulae beyond the Milky Way was unknown until Lord Rosse discovered it with his six-foot reflector in 1845. The Hooker telescope, greatly aided by optical and mechanical refinements and by the power of photography, can now record many thousands of these remarkable objects. Moreover, in the hands of Hubble it has shown that they are in fact "island universes," perhaps similar in structure to the Galaxy, of which our solar system is an infinitesimal part.

Our present instruments are thus powerful enough to give us this imposing picture of a universe dotted with isolated systems, some of them probably containing millions of stars brighter than our sun. It is also possible to measure the distance of the Great Nebula in Andromeda and one or two other spirals that lie about a million light-years from the earth. Much larger telescopes are needed, however, to continue the analysis of these nearest spirals, now only just begun, and to extend it to some of those at greater distances. Needless to say, the greater power of larger telescopes would also give us a far better understanding than we now possess of the structure and nature of the Galaxy, of which we still have much to learn. For example, we cannot yet say whether it shares the characteristic form of the spiral nebulae, nor do we even know with certainty whether it rotates about its center at the enormous velocity that seems equally characteristic of the "island universes." In fact, our own stellar system offers countless opportunities for productive research, as the important advances in our knowledge of the Galaxy recently made by Seares with the 60-inch Mount Wilson reflector so clearly indicate.

If our ideas of the structure of the universe are thus in a very early stage, the same may be said of our knowledge of the evolution of the stars. Recent discoveries in physics have greatly modified our conception of stellar evolution, affording a rational explanation of scores of questions formerly unanswered, but raising many new and fascinating problems. Giant stars with diameters several hundreds of times that of the sun, expanded by internal pressure to gossamer tenuity, lie near one end of our present stellar vista, with dwarfs of a density more than fifty thousand times that of water near the other. The sun, a condensing dwarf, 1.4 times as dense as water, stands on the downward slope of stellar life. The continual radiation that marks the transition from giant to dwarf is now attributed to the transformation of stellar mass into radiant energy, thus harmonizing with Einstein's views and accounting for the decrease in mass observed with advancing age. Surface temperatures ranging from about 3000° C. in the earlier stage of stellar life to about 100,000° at its climax, and internal temperatures perhaps reaching hundreds of millions of degrees are among the incidents of stellar existence. But here again, while theory and observation have recently joined in painting a new and surprising picture of celestial progress, important differences of opinion still exist and many of these await a more powerful telescope to discriminate between them. For while theories based on modern physics have been our chief guide in recent years, the final test is that of observation, and often our present instruments are insufficient to meet the demand.

So much in brief for the questions of celestial structure and evolution, though I have had to pass over the greatest of these problems: that of determining with certainty the successive stages in the development of the spiral nebulae, a phase of evolution vastly transcending that involved in the birth, life, and decline of a particular star. I have space to add only a word regarding the role of great telescopes in the study of the constitution of matter.

The range of mass, temperature, and density in the stars and nebulae is of course incomparably greater than the physicist can match in the laboratory. It is, therefore, not surprising that some of the most fundamental problems of modern physics have been answered by an appeal to experiments performed for us in these cosmic laboratories. For example, one of the most illuminating tests of Bohr's theory of the atom has just been made at the Norman Bridge Laboratory by Bowen in a study of the characteristic spectrum of the nebulae, where the extreme tenuity of the gas permits hydrogen and nitrogen to exist in a state harmonizing with the theory but unapproachable in any vacuum-tube. Similarly, Adams' observations of the companion of Sirius with the Hooker telescope confirmed Eddington's prediction that matter can exist thousands of times denser than any terrestrial substance. In fact, things have reached such a point that a far-sighted industrial leader, whose success may depend in the long run on a complete knowledge of the nature of matter and its transformations, would hardly be willing to be limited by the feeble range of terrestrial furnaces. I can easily conceive of such a man adding a great telescope to the equipment of a laboratory for industrial research if the information he needed could not be obtained from existing observatories.

The development of new methods and instruments of research is one of the most effective means of advancing science. In hundreds of cases the utilization of some obvious principle, long known but completely neglected, has suddenly multiplied the possibilities of the investigator by opening new highways into previously inaccessible territory. The telescope, the microscope, and the spectroscope are perhaps the most striking illustrations of this fact, but new devices are constantly being found, and the result has been a complete transformation of the astronomical observatory.

From our present point of view the chief question is the bearing of these developments on the design of telescopes. To Galileo a telescope was a slender tube, three or four feet in length, with a convex lens at one end for an object glass, and a concave lens at the other for an eyepiece. With this "optic glass" the surprising discoveries described in the *Sidereus Nuncius* were made, which shifted the sun from its traditional position as a satellite of the earth to the center of the solar system, and greatly enlarged the scale of the universe. After his time the telescope grew longer and longer, finally reaching the ungainly form of a lens supported on a pole as much as two or three hundred feet from the eyepiece. The invention of the achromatic lens brought the telescope back to manageable dimensions and permitted the use of an equatorial mounting, equipped with driving-clock to keep the celestial object at rest in the field of view. With the improvement of optical glass the aperture steadily increased, finally reaching 36 inches in the Lick and 40 inches in the Yerkes telescope.

Meanwhile it had become clear that the reflecting telescope, designed by Newton to avoid the defects of single lenses, possessed many advantages over the refractor. Chief among these are its power of concentrating light of all colors at the same focus and the fact that the light does not pass through the mirror, but is reflected from its concave front surface. Speculum metal, a highly polished alloy of tin and copper, was used for the early reflectors, reaching a maximum size in Lord Rosse's six-foot telescope. Mirrors of glass, silvered on the front surface, were then introduced, and proved greatly superior in lightness and reflecting power. Moreover, optical glass perfect enough for lenses cannot be obtained in very large sizes, and even if it could, the loss of light by absorption in transmission through the glass would prevent its use for objectives materially exceeding that of the Yerkes telescope. Therefore, our hopes for the future must lie in some form of reflector.

It is evident that a lens, through which the starlight passes to the eye, must be mounted in a very different way from a concave mirror, which receives the light on its surface and reflects it back to the focus. The large concave mirror lies at the bottom of the telescope tube, which is usually of light skeleton construction, open at the top. The surface of the mirror is figured to a paraboloidal form, which differs somewhat from a sphere in curvature, and has the power of concentrating the parallel rays from a star in a point at the focus. This focus is near the top of the tube, opposite the center of the mirror.

For some classes of work it is desirable to place the photographic plate, small spectroscope, or other accessory instrument at this principal focus, centrally within the tube. Some starlight is thus cut off from the large mirror, but the loss is small and is less than with other arrangements. Newton interposed a plane mirror, fixed at an angle of 45°, which reflected the light to the side of the tube, where he placed the eyepiece. Cassegrain substituted a convex mirror for Newton's plane. Supported centrally at right angles to the beam, it changes the convergence of the rays and brings them to a focus near the large mirror. An inclined plane mirror may be used to intercept them, thus bringing the secondary focus at the side of the tube, or the large mirror may be pierced with a hole, allowing the rays to come to a focus close behind it.

In a third arrangement, the rays may be sent through the hollow polar axis of the telescope to a secondary focus at a fixed point in a constant temperature laboratory. This arrangement, first suggested by Ranyard, was embodied with both the Newtonian and Cassegrain methods in the mountings of the 60-inch and 100-inch telescopes of the Mount Wilson Observatory. By these means we may obtain any desired equivalent focal length (which varies with the curvature and position of the small convex mirrors) and thus photograph celestial objects on a large or small scale, as required by the problem in hand. Furthermore, we can use to the best advantage all types of spectroscope, photometer, interferometer, thermocouple, radiometer, photo-electric cell, and the many other accessories developed in recent years.

These accessory instruments and devices have made possible most of the discoveries of modern astrophysics. The stellar spectroscope, originally merely a small laboratory instrument attached to a telescope, has grown to the dimensions of the powerful fixed spectrograph of 6 inches aperture and 15 feet in length, recently used with splendid success by Adams in photographing the spectra of some of the brightest stars. The development of this method of high dispersion stellar spectroscopy, initiated in the early days of the Yerkes Observatory, was one of my chief incentives in endeavoring to obtain large reflecting telescopes for the Mount Wilson Observatory. The recent advances in our knowledge of the atom and the consequent complete transformation of spectroscopy from an empirical to a rational basis greatly increase the possibilities of analyzing starlight. In most of the small-scale spectra photographed with ordinary stellar spectrographs the lines are so closely crowded together that they cannot be separately measured. With a larger telescope we could push the dispersion to the point attained by Rowland in his classic studies of the solar spectrum, and thus take full advantage of the great possibilities of discovery offered us by recent advances in physics.

These details are important because they point directly to the type of telescope required. It is true that in some cases lenses may be used instead of convex mirrors for enlarging the image; but in our judgment the design should permit observations to be made in the principal focus of the large mirror, at a secondary focus just below the (pierced) mirror, and at another secondary focus in a fixed laboratory. It should also be possible to attach to the tube a large Michelson stellar interferometer, arranged for rotation in position angle and thus suitable for the measurement of very close double stars.

A mounting designed by Pease of the Mount Wilson Observatory meets these requirements and is worthy of careful consideration. It is large enough to carry a mirror 25 feet in diameter, collecting nine times as much light as the 100-inch Hooker telescope. It would thus enlarge our sphere of observation to three times its present diameter and increase the total number of galactic stars to three or four times that now within range.

This, of course, is a tentative design, subject to modification in the light of an exhaustive study. Of all the optical and mechanical problems involved only one presents real difficulties, but there is no reason to think that these cannot be readily surmounted. This is the manufacture of the glass for the large mirror.

Our chief difficulty in the case of the Hooker telescope was to obtain a suitable glass disc. The largest previously cast was that for the 60-inch mirror of our first large reflector. This is 8 inches thick and weighs a ton. The 100-inch disc, 13 inches thick, weighs nearly five tons. To make it three pots of glass were poured in quick succession into the mold. After a long annealing process, to prevent the internal strains that result from rapid cooling, the glass was delivered to us. Unlike the discs previously sent by the French makers, it contained sheets of bubbles, doubtless due in part to the use of the three pots of glass, while but one had sufficed before. Any considerable lack of homogeneity would result in unequal expansion or contraction under temperature changes, and experiments were, therefore, continued at the glass factory in the Forest of St. Gobain in the hope of producing a flawless disc. As they did not succeed, the disc containing the bubbles was given a spherical figure and tested optically under a wide range of temperature. Its performance convinced us that the disc could safely be given a paraboloidal figure for use in the telescope, where it has served ever since for a great variety of visual and photographic observations.

Recently, important advances have been made in the art of glass manufacture, and mirror discs much larger and better than the 100-inch can now undoubtedly be cast. Pyrex glass, so useful in the kitchen and the chemical laboratory because it is not easily cracked by heat, is also very advantageous for telescope mirrors. Observations must always be made through the widely opened shutter of the dome, at temperatures as nearly as possible the same as that of the outer air. As the temperature rises or falls the mirror must respond. The small expansion or contraction of Pyrex glass means that mirrors made of it undergo less change of figure and, therefore, give more sharply defined star images — a vitally important matter in all classes of work, especially in the study of the extremely faint stars in the spiral nebulae, for which Pease's design is especially adapted.

Dr. Arthur L. Day of the Carnegie Institution of Washington, working in association with the Corning Glass Company, has succeeded in producing glass with a higher silica content than Pyrex and, therefore, with a lower coefficient of expansion. Moreover, Dr. Elihu Thomson and Mr. Edward R. Berry of the General Electric Company have recently made discs up to 12 inches in diameter of transparent fused quartz (pure silica), which is superior to all other substances for telescope mirrors. The chief difficulty in the manufacture of fused quartz has been the elimination of bubbles. These would do no harm whatever within a large telescope mirror, provided its upper surface were freed from them by a method proposed by Dr. Thomson. In fact, the presence of a great number of bubbles would be a distinct advantage in reducing the weight of the disc. As there is every reason to believe that a suitable Pyrex or quartz disc could be successfully cast and annealed, and as the optical and engineering problems of figuring, mounting, and housing it present no serious difficulties, I believe that a 200-inch or even a 300-inch telescope could now be built and used to the great advantage of astronomy.

Limitations of space have prevented mention of many interesting matters of detail. It goes without saying that all questions relating to the optical as well as the engineering design should be thoroughly investigated by a group of competent authorities, who should also include those best qualified to deal with related problems involving the design of spectroscopes and the many other accessory instruments required. As for photographic plates, it is well known that the power of photographic telescopes could be materially increased by improving their quality, so that no effort in this direction should be spared.

Perhaps a word as to procedure may be added. The first step should be to determine by experiment how large a mirror disc, preferably of fused quartz, can be successfully cast and annealed. Meanwhile all questions as to the final design of the mounting and accessories could be settled. With the completion of the mirror disc the only uncertainty would vanish and the optical and mechanical work could begin.

*O*n the evening of May 15, 1964, Dr. Ira S. Bowen, then Director of the Mount Wilson and Palomar Observatories, delivered an address before the Trustees, staff and friends of the Carnegie Institution in Washington, D. C., evaluating the first fifteen years of observations by the 200-inch Hale telescope at Palomar. This article, subsequently prepared by Dr. Bowen for Science, was based on that address.

Born at Seneca Falls, New York, in 1898, Dr. Bowen has devoted his entire career to the advancement of our knowledge of the universe: as administrator, as observer, and as designer of astronomical instruments. After receiving his A.B. degree from Oberlin College in 1919, he began graduate studies at the University of Chicago, where he had the good fortune to meet Dr. Robert A. Millikan, Chairman of the Executive Council of California Institute of of Technology. At Dr. Millikan's invitation, he continued his graduate work at Caltech, receiving his Ph.D. degree in 1926. He remained at Caltech until 1945 as a professor of physics. During the same period he did pioneering work in the then new field of spectroscopy.

When the 200-inch Hale telescope project began to take form in the 1930s he early became associated with it. He designed the coudé spectrograph for the telescope and was responsible for the final testing and finishing of the great mirror. He was named Director of Mount Wilson Observatory in 1946, and two years later, when the two observatories were combined, under the joint operation of Carnegie Institution and Caltech, he became Director of Mount Wilson and Palomar Observatories. As Director he was a major contributor to the success of the great observational programs that provided so many new insights into the nature and structure of the universe in the subsequent fifteen years.

Dr. Bowen retired from the Directorship of the Observatories on June 30, 1964, after eighteen years with Carnegie Institution. He is continuing his work on various aspects of design and improvement of large telescopes as a Distinguished Service Member of the Institution.

250

Ira S. Bowen

EXPLORATIONS WITH THE HALE TELESCOPE

From *Science,* Vol. 145, pp. 1391–1398, September 25, 1964. Copyright 1964 by the American Association for the Advancement of Science.

In the April 1928 number of *Harper's Magazine* there appeared an article by George E. Hale entitled "The Possibilities of Large Telescopes." This article came to the attention of Wickliffe Rose of the International Education Board, and this board, in cooperation with the General Education Board, provided the funds for the construction of the Hale 200-inch telescope on Palomar Mountain.

In his article Hale wrote, "Other reasons that combine to assure the success of a large telescope are the remarkable opportunities for new discoveries revealed by recent astronomical progress and the equally remarkable means of interpreting them afforded by recent advances in physics. These new possibilities are so numerous that I must confine myself to three general examples, bearing upon the structure of the universe, the evolution of stars, and the constitution of matter."

It is now nearly a decade and a half since the Hale telescope began regular observations. We may therefore appropriately ask to what extent the opportunities for discoveries in the fields enumerated by Hale have been realized by the 200-inch telescope and its supporting instruments.

251

STRUCTURE OF THE UNIVERSE

Consider first the structure of the universe. When Hale wrote in 1928, Hubble had just succeeded in resolving the stars in the Andromeda galaxy with the 100-inch telescope on Mount Wilson. By identifying among these stars a number of cepheid variables and by comparing their apparent brightness with the absolute brightness, as determined from examples in our Milky Way, he arrived at a distance for Andromeda of a little less than 1 million light-years. Similar comparisons, in which, finally, the brightest galaxies in a cluster of galaxies were used as a distance indicator, led Hubble to an estimate of 500 million light-years as the maximum distance observable with the 100-inch.

All these observations were at the extreme limit of the capabilities of the 100-inch, and it was realized that all the measurements were subject to large uncertainties. One of the high-priority programs planned for the Hale telescope was a thorough reexamination of all the steps in determining the distance of Andromeda and more distant objects. First, a recalibration of the whole scale of stellar magnitudes was undertaken, by means of the new and more precise photoelectric techniques. In the hands first of Stebbins and Whitford and later of Baum, these remeasurements showed that the older photographically determined magnitudes were more and more in error as fainter magnitudes were reached until, in the range from the 20th to the 23rd magnitude, which had been used in the Andromeda studies, the error approached a full magnitude.

Next, a reevaluation of the absolute magnitudes of nearby cepheids as a function of period was undertaken by Baade at Palomar, by Thackeray and Wesselink at Pretoria, and by Mineur, Blaauw, and H. R. Morgan. All these groups arrived independently at the conclusion that the classical cepheids are about 1.5 magnitudes brighter than had been supposed.

Finally, Baade carried out a very extensive investigation of the apparent magnitudes of the cepheid variables in Andromeda. To establish light curves of these variables it was necessary to obtain, with the 200-inch, well over 100 photographs of each of the four fields under study. Areas I, II, and III were photographed first, but a preliminary study of the variables gave erratic magnitudes, indicating that many of the cepheids were embedded in dust clouds that absorbed part of their light. Field IV, far out from the nucleus of Andromeda, was then photographed. Here the variables are much fewer but are free of obscuration. More than 300 cepheids and numerous other variables were located by Baade on the plates of the four fields. Henrietta Swope then measured the magnitude of each of the cepheids or other variables in fields I, II, and IV, on each of the 100 or more plates on which the star was located, and drew a light curve in each of two colors.

Unfortunately Baade did not live to see the completion of these measurements. Swope's final analysis, which has just been completed, when considered in combination with the new magnitude scale and the new values for the luminosity of the cepheids, leads to a distance for Andromeda of 2.2 million light-years, or three times Hubble's value.

This new value also means that Andromeda has a diameter 3 times, a luminosity 9 times, and a mass 3 times the earlier estimates. Hubble's original values indicated that Andromeda and other large galaxies were all appreciably smaller than our Milky Way and led one well-known astronomer to remark in the 1920's, "If we call them islands, the Galaxy is a continent." Swope's latest value for the distance indicates that Andromeda is both larger and more massive than the Milky Way.

The extension of these and related methods to the study of more distant galaxies is still in progress, and final values for distances and related constants are not yet available. Preliminary estimates by Sandage and others, however, indicate that Hubble's values for the distances of the farthest galaxies he observed must be increased by an even greater factor, of perhaps 4 to 7.

Following Hubble's identification of the galaxies as distant stellar systems, Hubble and Humason investigated the spectra of these objects. This led to the discovery that all their spectra are shifted toward the red end of the spectrum by an amount that is approximately proportional to the distance. Interpreted as a velocity of recession, the red shift led to the concept of the expanding universe, which has been fundamental to all cosmological theories since that time. In the original study with the 100-inch, Humason observed galaxies with red shifts of up to 13 percent of the wavelength. With the completion of the 200-inch telescope he was able to observe galaxies with shifts of more than 20 percent.

RADIO SOURCES AND GALAXIES

In the meantime a new and unexpected approach to many of these problems developed from the discovery that, in addition to light, short-wave radio waves are coming from outside the atmosphere. Much of the radiation comes from huge clouds of gas in the spiral arms of our Galaxy and has proved a very powerful tool in outlining the position of the arms. More detailed study revealed a large number of small localized sources. As radio techniques improved it was possible to locate the sources with sufficient accuracy to identify them with optically observed objects. Minkowski and Baade in the mid and late 1950's were especially successful in making identifications on plates made with the 48-inch Schmidt and the 200-inch Hale telescopes. A number of the sources were found to be objects in the Milky Way, several being remnants of supernova explosions, including those recorded in A.D. 1054, 1572, and 1604.

A number of additional radio sources were identified with galaxies. All the radio galaxies close enough for detailed study (for example, NGC 5128) show many peculiarities. Other galaxies, such as M87, show jets of matter being ejected. In general the light of the jets is strongly polarized, indicating that much of the light comes from very-high-speed charged particles (electrons or protons) being accelerated in a magnetic field, the so-called synchrotron radiation. Spectroscopic studies show strong emission lines of abnormal strength. Furthermore, the lines are wide and often have complicated structures indicating large relative motions of a thousand or more kilometers per second between various parts of the galaxy. Because of the dual character of several of the galaxies and the high relative velocities of their parts, several of these objects were originally interpreted as galaxies in collision.

Four years ago Minkowski reported on the identification of one such radio source, 3C295, with the brightest member of a very faint cluster of galaxies. Like the spectra of most radio sources, its spectrum was characterized by strong emission. The red shift was the largest that had been observed — 46 percent of the wavelength. This red shift was confirmed by Baum from the shift of the maximum of the continuous radiation of two of the normal galaxies in the same cluster. These observations placed the cluster at the greatest distance of any object then known.

Less than 2 years ago Matthews (from California Institute of Technology Radio Observatory) and Sandage identified three of the radio sources with stellar objects, although one or two showed faint wisps of nebulosity extending out from them. Spectrograms taken with the 200-inch showed emission lines, which, however, did not agree in position with any known lines. Finally, a detailed study of the spectra by Schmidt, in combination with infrared scanner observations by Oke, gave a convincing identification of the lines in the object 3C273 with well-known nebular lines shifted toward the red by 16 percent from their normal positions. Following this clue, Greenstein and Matthews were able to interpret their spectra of 3C48 in a similar way, finding a shift toward the red of 37 percent. A theoretical discussion by Greenstein and Schmidt has shown that the red shift can be interpreted best as a velocity of recession similar to that characteristic of all distant galaxies. Interpretation on the basis of the usual Hubble relationships between distance and red shift places these objects at such a distance that their absolute brightness is nearly 100 times that of any normal galaxy such as Andromeda.

At about the same time the very detailed study of these radio galaxies made with the California Institute of Technology radio interferometer in Owens Valley showed that the radio sources associated with these objects are often double, the two lobes being placed symmetrically on opposite sides of the optical object and at distances of a few tens or hundreds of thousands of light-years from it.

Sandage, in collaboration with Lynds of the Kitt Peak Observatory, then investigated M 82, one of the nearby radio galaxies. They found from spectroscopic observations that the streamers moving out from the nucleus on both sides of the galaxy had at each point a velocity proportional to the distance from the nucleus. This, of course, indicates that all parts of the material left the nucleus at the same time, about a million and a half years ago.

All these observations point to some enormous release of energy which can occur in the nucleus of a galaxy and which can cause the galaxy to emit energy in amounts as large as 100 times the normal radiation from all the stars of a large galaxy such as Andromeda. At the same time, great numbers of charged particles of very high energy are ejected normal to the plane of the galaxy, giving rise to the pair of radio sources, and large quantities of matter are thrown off, producing the effects studied by Sandage and Lynds. Obviously an entirely new mechanism of tremendous power has been discovered. Present indications are that a substantial fraction of the several thousands of known radio sources are galaxies undergoing an explosion of this type. This is therefore not a rare phenomenon, particularly when we consider that the lifetime of the explosion, perhaps a few million years, is very short in comparison with the life of a galaxy. Although it is too early to propose any definite theory about the exact mechanisms involved, obviously these events must play a very major role in the evolution of many galaxies.

IMPLICATIONS FOR COSMOLOGY

The objects are also of great importance for the study of cosmology. Since they are up to 100 times as bright as any normal galaxy they can be observed at a much greater distance. Furthermore, owing to their intense activity, these sources emit a large amount of radiation in the far-ultraviolet, especially in the form of emission lines. At great distances the radiation is red-shifted up into the easily observable range. Thus is eliminated the difficulty that has set a limit on observations of normal galaxies — namely, that the radiations, which are chiefly in the visual region, are shifted at these great distances far into the infrared, out of the range of sensitive receivers.

For example, during the past few months Matthews and Schmidt have observed a red shift of 54.5 percent in the spectrum of the radio galaxy 3C147. This makes 3C147 the most distant object thus far located. Indeed, its distance is so large a fraction of the radius of the universe that corrections which depend on the cosmological model of the universe adopted are large and uncertain. These uncertainties, in combination with the uncertainties that remain in the distance scale, make it impractical to quote a definite distance in light-years. It is abundantly clear, however, that we are observing with the 200-inch at distances greater by a whole order of magnitude than Hubble's 1940 values for the limit reached with the 100-inch. Thus, owing to the much greater penetration in space of the 200-inch and the several-fold increase in the distance scale, we now discuss these distances in billions of light-years, whereas Hubble listed his most distant objects in hundreds of millions of light-years.

Of far greater significance than the increase in distance is the fact that, in extending the observations to a large fraction of the radius of the universe, we have reached the region where it should be possible to differentiate observationally among the various cosmological models.

STELLAR EVOLUTION

The second and third of Hale's examples of problems for the 200-inch were the evolution of stars and the constitution of matter. The two programs have developed together, and the advances in one field have provided the keys to the problems of the other.

On the observation side, the first major step in the investigation of stellar evolution was taken by Baade in his observations of the Andromeda galaxy from Mount Wilson during World War II. Thanks to the blackout in Los Angeles, the skies above Mount Wilson were dark, and Baade was able to make a long series of photographs of the galaxy in light of different colors. He noted that in the spiral arms the brightest stars are blue, and that they are accompanied by extended luminous clouds of gas. In the nuclear region, however, the gas clouds are missing and the brightest stars are red.

Baade correctly guessed that the bright blue giants in the spiral arms have recently condensed from the gas clouds and hence are very young stars, and that the red giants in the nuclear regions are old stars. He designated the young stars as population I, since they have characteristics similar to those of the stars in the neighborhood of the sun, which is located in one of the spiral arms of the Milky Way. He called the old stars population II.

At the time Hale wrote his article the mechanism of

the production of the enormous energy radiated by the stars was not known, and, of necessity, theories about the evolutionary development of a star were the crudest speculations. By the late 1930's, however, advances in nuclear physics had made it clear that the primary source of the energy is the transformation of hydrogen into helium in the hot core of the star. On this basis theories of stellar structure were developed which showed that, if a large mass of gas, chiefly hydrogen, condenses into stars of various masses, certain relationships must hold between the luminosity, the surface temperature, and the mass. The relationships are shown in a curve at left in Fig. 1, where the luminosity is plotted against the surface temperature, as indicated by the color. The point on the curve where a given star is located is fixed by its mass, the luminosity being proportional to about the third power of the mass.

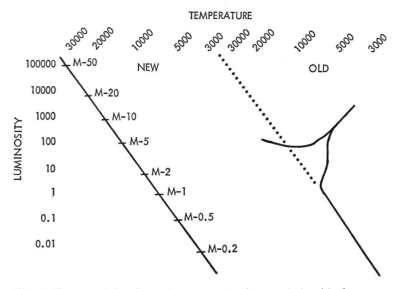

Fig. 1. Curve at left (the main sequence) shows relationship between luminosity, surface temperature, and mass of stars while burning hydrogen. Curve at right shows the relationships after the more luminous stars have exhausted their hydrogen fuel.

Until the hydrogen fuel in the stellar core drops to a certain critical value the stars continue to have the properties shown by this curve, which is known as the main sequence. Since the rate of radiation goes up much more rapidly than the mass of available fuel, the very bright massive stars use up their hydrogen much sooner than the smaller stars do. The theory of stellar interiors predicts that when the fuel is exhausted the surface of the star will cool off but the star, at the same time, will expand so much that its brightness will increase. In other words, in the diagram (Fig. 1) the star moves off the main sequence up and to the right. The star remains in the expanded red stage for a relatively short period, then its luminosity drops and its surface temperature rises; it often passes through a stage of rapid fluctuation in brightness, with a period of a day or less. Finally the star ends up as an exceedingly dense white dwarf in the region to the left of the main sequence.

An old group of stars would therefore have the distribution shown at the right in Fig. 1. Here the brightest stars are red, corresponding to Baade's population II. Furthermore, as the group of stars grows still older the point of break-off from the main sequence moves down lower on the diagram. If, therefore, the magnitudes and colors of a homogeneous group of stars are measured and plotted on such a diagram, the position of the break-off point indicates the interval since the original condensation of the stars occurred, and thus the age of the group. The development of precise photoelectric photometry shortly after the war made this type of measurement feasible.

To establish a precise curve it is necessary to measure the magnitude in two or more colors for each of a few hundred stars in the group or cluster of stars under investigation. During the past 15 years several scores of globular and galactic clusters and other groups of stars have been observed in detail by Sandage, Baum, Arp, Eggen, and others. A comprehensive pattern of the ages of various parts of our Galaxy and of several nearby galaxies has been developed. Ages from a million years for the youngest galactic clusters up to 12 billion years for the oldest globular clusters have been found. The ages are in satisfactory agreement with the age of the universe as estimated from cosmological investigations. The age of 5 billion years for the sun and solar system, as calculated by other procedures, fits this pattern satisfactorily.

A very effective joint attack on the problems of elucidating the detailed mechanisms by which the stellar evolution occurs, especially in its later stages, and the simultaneous changes that occur in the chemical composition of the star as it grows older, has been made through very close cooperation between the nuclear physicists at the California Institute of Technology and the Observatory staff. Shortly after the war William Fowler of the Institute physics department became interested in the nuclear transformations that may occur under the conditions of temperature and pressure that exist in stellar cores. The experiments and theoretical studies of Fowler and his collaborators have provided much of the physical basis for the theories of stellar evolution. In a parallel large project supported by the Air Force Office of Scientific Research, Greenstein and his collaborators have used the very fast and efficient spectrographs of the 100-inch and 200-inch telescopes to make detailed quantitative chemical analyses of large numbers of stars of different ages and evolutionary histories. From the studies the following picture of the evolutionary history of a star has been developed.

The gas clouds from which the star condenses are made up chiefly of hydrogen. As the mass of hydrogen condenses into the star its core is heated to a temperature of the order of 10^7 degrees Kelvin. At this temperature hydrogen is slowly transformed into helium by one or both of two possible mechanisms. Each kilogram of hydrogen transformed into helium produces an amount of energy equivalent to the combustion of about 20,000 metric tons of the best coal. While this transformation of hydrogen continues the star remains stable, with properties corresponding to those of stars on the main sequence. When the hydrogen fuel in the core approaches exhaustion, the star moves off the main sequence and the core heats to about 100 million degrees. At these temperatures the helium atoms, which are now the chief constituent of the core, can react to form carbon, nitrogen, oxygen, and neon. These reactions also liberate a number of neutrons which can combine with the atoms present to form the heavier elements, such as iron. As the reactions continue, the core temperature may eventually increase to a few billion degrees. In the more massive stars the reactions, because of the star's gravitational instability, may eventually proceed explosively, being completed in a matter of hours or even seconds. This is presumably the cause of supernovae, in which, for the period of a few weeks after the explosion, the star emits as much light as all the billions of normal stars in a whole galaxy. During the explosion a substantial fraction of the mass of the star is thrown off into space with a velocity of thousands of kilometers per second. In stars of smaller mass these reactions normally proceed more slowly, taking millions of years to reach completion. Even in these smaller stars, as Deutsch has shown, some of the material of the star is slowly blown off into space.

STELLAR COMPOSITION

The spectroscopic observations show that all stars contain a small amount of the heavier elements, like calcium and iron. Since the heavy elements cannot be formed in the star until long after it has moved off the main sequence, these elements must have been present as an impurity in the hydrogen clouds that condensed into the star. Recent quantitative measurements do show, however, that the heavy metals are much more abundant, often by a factor of 100 or more, in young or moderate-age stars, like the sun, than they are in very old stars, such as those in certain globular clusters that condensed very soon after the formation of the galaxy. Presumably part of the material that condensed into the younger stars had already passed through one or more earlier generations of stars, during which the metals were formed, and was then blown off into space to mix with the uncondensed hydrogen.

In addition to these gross variations in the ratio of all heavy elements to hydrogen, many anomalies in the abundances of individual elements such as lithium, beryllium, carbon, nitrogen, and phosphorus have been observed and measured in a number of peculiar stars. The mechanisms that might have produced such anomalous abundances have been investigated in great detail by Fowler's group.

The supernovae explosions have been another subject of extensive study. In one project, under the supervision of Zwicky, a monthly patrol for discovery of the outbursts has been maintained. Several score supernovae have been found, and their light curves and spectra have been followed by various observers.

OTHER INVESTIGATIONS

As was predicted by Hale in 1928, these have been the three major fields of research with the 200-inch telescope. Investigations have also been made in numerous more specialized fields; limitations of space permit mention of only two or three.

One of Hale's most important discoveries in astronomy was the finding of localized magnetic fields in sunspots — the first time magnetic fields had been observed outside the earth. In the late 1940's Horace Babcock initiated a search for magnetic fields in other stars. Large and often variable magnetic fields that evidently covered a substantial fraction of the stars' surface were discovered. More than a third of the several hundred stars that have now been examined show evidence of such fields. In one, a magnetic field of more than 34,000 gauss was measured.

Finally, closer to home, a number of important observations have been made of bodies in the solar system. During the Mariner II flyby of Venus in December 1962, Murray made infrared observations with the 200-inch paralleling those obtained from Mariner II. The results secured with the telescope were much more extensive than those obtained from the spacecraft and on one night showed a point of very intense storm activity in the atmosphere of Venus. A year ago Münch, with the assistance of Spinrad and Kaplan of the Jet Propulsion Laboratory, used the 100-inch to make the first positive observation and measurement of the water vapor in the atmosphere of Mars. At the same time these workers observed faint bands of carbon dioxide in the near-infrared. From the strength of these lines relative to stronger lines in the far-infrared, observed earlier by Kuiper, they were able to determine that the total pressure of Mars' present atmosphere is between 1/5 and 1/2 that of earlier estimates. This observation has made necessary a major modification of plans to parachute equipment to the Martian surface.

The 200-inch telescope has performed much as Hale had hoped and predicted. Most of the programs he listed have made large advances. A very few have run into unforeseen difficulties. On the other hand, important breakthroughs have occurred along lines of which Hale had no inkling. If Hale were with us today I believe he would be content with the results of the great adventure he started.

*I*n *"The Big Schmidt"* Dr. Albert Wilson gives the history and anatomy of an increasingly famous scientific instrument, the 48-inch Schmidt telescope of the Palomar Observatory in California. The article appeared in the December 1950 issue of Scientific American.

The Schmidt telescope is indeed well known to Dr. Wilson; he has used it for a variety of research projects, including the famous sky survey sponsored by Palomar Observatory and the National Geographic Society. Dr. Wilson was the astronomer in charge of that four-year mapping of the skies.

Dr. Wilson has also carried out visual and photographic studies with the 18-inch Schmidt. With Fritz Zwicky and Josef Johnson of the Carnegie Institution, he made a survey of our Milky Way galaxy; and in 1947 Wilson and J. C. Duncan, using the little Schmidt, found 12 new planetary nebulae.

Albert G. Wilson was born in Houston, Texas, in 1918. He was graduated from Rice Institute in 1941, and then attended the California Institute of Technology for advanced study in mathematics. He received his master's degree in 1942 and his doctor's degree in 1947.

For the next two years Wilson was senior research fellow in astrophysics at California Institute of Technology. From 1949 to 1953 he was a staff member of the Carnegie Institution at Mount Wilson and Palomar Observatories, which are jointly operated by Carnegie and Caltech. In 1953 Wilson left Carnegie for the Lowell Observatory, where he served as assistant director from 1953 to 1954 and as director from 1954 to 1957. Since 1957 he has been with the Rand Corporation.

Dr. Wilson is a Fellow of the Royal Society and a member of the Astronomical Union and the American Mathematical Society.

266

Albert G. Wilson

THE BIG SCHMIDT

From *Scientific American*, December 1950.

There was a time when the only instruments an astronomer needed in the practice of his profession were a conventional telescope and a good clock. Today the problems astronomy is tackling are so intricate and diverse that an astronomer who set out to investigate them armed only with a traditional telescope would be a little like a naturalist setting out to collect whales and microbes with a butterfly net. The modern astronomer, concerned with such matters as the method of energy generation in stars, the relative abundances of the chemical elements in the universe and the distribution in space of fantastically far-off galaxies, requires a large array of special instruments, each designed for a particular purpose. His equipment nowadays includes such devices as photo-electric photometers, radio "telescopes," high-altitude rockets and devices for producing artificial eclipses of the sun. And the evolution of the optical telescope itself has produced a number of highly specialized forms.

Public attention has focused mainly on Palomar Mountain's giant 200-inch telescope. Not many people realize that the 200-inch is a specialist of a kind. It looks a billion light-years into space, but it gives us only a gimlet-eyed view; what it gains in penetration it loses in breadth of vision. Looking through such a telescope is like looking into a ball park through a nail hole in the fence.

The 200-inch is the culmination of the main line of development in classical telescopes. We have been building ever larger and more powerful instruments, reaching farther and farther into space but steadily narrowing our view. As a result we have in a sense been seeing less and less of the sky as a whole. At the farthest range of our more powerful telescopes less than two per cent of the sky has been photographed so far. Our present picture of the universe is based on a few long thin views of these remote regions and on what we know about the regions near us.

Obviously to get a comprehensive view of the universe we need a new type of telescope that can look both far and wide at the same time. The best answer to this need has been found in the large Schmidt-type photographic telescope. By an ingenious combination of mirror and lens it can photograph the sky at once to great depths and over a wide field. This article will describe the largest instrument of this type now in existence — the 48-inch Schmidt on Palomar Mountain.

ABERRATIONS

The main problem in designing a telescope to provide a wide field of view is to get rid of the optical aberrations or interferences with the quality of the image that are inherent in reflectors and refractors. One of these aberrations goes by the name of chromatism. It occurs in all lenses, and it is caused by the fact that the refracting lens splits the transmitted light slightly into its spectrum of wavelengths, thereby producing a colored fringe that makes the image fuzzy. Another fault, common to both lenses and mirrors, is spherical aberration, arising from the fact that the different zones of even a perfect spherical lens or mirror focus the light falling on them at different points; the result again is a hazy image. Other serious aberrations are astigmatism and coma. These defects, which affect only off-axis rays, are caused by unequal magnification of the different zones and become more serious as the angle of the ray with the axis increases, *i.e.*, toward the edges of the picture.

The history of telescopes is largely a history of the various devices used to remove aberrations. In the early telescopes, which were mainly refractors, chromatism was mitigated by using lenses with a very large focal length; this reduces the effects of differences in refraction of the various wavelengths. A telescope with a large focal length is, however, unwieldy. Later it was discovered that two or more lenses with different refractive indices could be used in combination to correct the color defects.

The largest modern telescopes are all reflectors, so chromatism is not a problem. But the large reflectors must still contend with the other aberrations. Spherical aberration is usually overcome by making the curve of the primary mirror a parabola rather than truly spherical. A parabolic reflector, however, possesses the off-axis aberrations, and it is these, principally the coma, which cause the trouble today. They are the principal reason why the modern large reflecting telescopes, though mighty in light-gathering power and ability to penetrate to great depths of space, have so narrow a field.

Several solutions have been proposed for the removal of the off-axis defects. Small gains in the size of the field can be obtained by introducing a correcting lens near the focus; in the 200-inch such a lens increases the usable field from 2 minutes of arc to about 15 minutes (a quarter of a degree) of arc. Other systems, using two mirrors to free images of spherical aberration and coma, have made possible still larger fields. But the most radical and also the most successful design is that of the late Bernhard Schmidt of the Hamburg-Bergedorf Observatory in Germany.

SCHMIDT'S IDEA

To rid the optical system of off-axis aberrations, Schimdt resorted to a revolutionary remedy; he did away with the axis. He decided to use a spherical mirror, because a sphere has no axis and no off-axis aberrations, and to try to solve the problem of spherical aberrations in a new way. He conceived the idea of altering the direction of the incident rays *before* they reached the spherical mirror in such a manner that after reflection they would be brought to the same focus by all zones. To effect this Schmidt designed a lens, properly called a correcting plate, which he placed out in front at the center of the curvature of the mirror. This combination lens-mirror design removes both spherical aberration and coma and provides images of excellent definition over fields several degrees in diameter. Schmidt's system has the further advantage that it can be built with photographic speeds faster than $f/1$.

On the other hand, there are some disadvantages to the Schmidt system. One is that the picture is not focused on a flat plane but on a spherical surface. This means that the photographic plate must be bent to a spherical surface or that a third optical element must be introduced into the system to flatten the field. Further, the focal surface, where the photographic plate must be placed, is located about halfway between the mirror and the correcting plate — a not readily accessible position. In large Schmidts the problem of loading and unloading the plates can become a serious one. But all of these disadvantages are of little weight when compared with the great advantages of excellent definition, wide field and high speed.

It was the amateur astronomers who did much of the pioneering work with Schmidt cameras. Partly because of their valuable experiments on a small scale, it was soon realized that a Schmidt camera of large dimensions would be the ideal answer to the need for a wide-angle camera capable of photographing faint objects.

How large a camera of this type was it feasible to build? Schmidt's first camera, built in 1930, was an $f/1.74$ system with a 14-inch correcting plate and a 17-inch mirror. North American observatories began to construct larger and larger models: the Palomar Observatory had one built with an 18-inch correcting plate and a 26-inch mirror (18/26), the Warner and Swasey Observatory in Cleveland a 24/36, the Harvard College Observatory a 24/33, the Mexican National Astrophysical Observatory a 26/30. But it seemed that Schmidts much beyond this size might not be feasible.

There was no problem about getting a large mirror; successful mirrors with diameters up to 100 inches had been built. The trouble lay in increasing the size of the correcting plate. The Schmidt correcting plate, unlike a conventional lens, is very thin; on the average it has a thickness of the order of only one fiftieth of its diameter. Consequently as it is increased in size the elastic bending of the thin plate may become appreciable. Fortunately, however, the optical system is relatively insensitive to such deflections, and it is even possible to support the correcting plate at its center, so this factor is not a serious limitation on the plate's size. More serious is the fact that the correcting plate, like all lenses, is subject to chromatic aberration. It is possible, of course, to achromatize the system by employing two plates of different indices of refraction. But in large sizes this might be very difficult.

THE BUILDING OF THE 48-INCH

In 1938 the Observatory Council of the California Institute of Technology, the group responsible for the design and construction of the 200-inch telescope and its auxiliaries, decided that a large Schmidt-type camera would make an excellent auxiliary for the 200-inch. It was decided to build as large a Schmidt as could feasibly be constructed without the necessity of revolutionary modifications to overcome chromatism. Calculations showed that an $f/2.5$ camera with a 48-inch conventional correcting plate and 72-inch mirror would not introduce objectionable chromatism. In 1939 construction was begun on such a Schmidt for the Palomar Observatory.

Overshadowed by its giant colleague, the 48-inch Schmidt attracted little attention. But its engineering and optics required the same sort of highly skilled techniques that were demanded for the 200-inch. The most exacting single item was the shaping of the large correcting plate. This pioneering piece of work was taken over by Don Hendrix of the Mount Wilson Observatory optical staff. Hendrix inspected a whole carload of plate glass before he found a piece sufficiently free of defects to make a useful blank for the correcting plate. He then devised tools and methods for grinding, polishing and testing the large plate and in only three months' working time succeeded in producing a correcting plate of excellent quality. Hendrix also shaped the spherical surface for the 72-inch Pyrex mirror. The remaining optical parts, including two 10-inch refractors used for guide telescopes, were made in the optical shops of the California Institute of Technology.

The telescope itself was manufactured in the Caltech machine shops and set up on Palomar in a dome about a quarter of a mile east of the 200-inch. The Big Schmidt, as it is called, consists of a tube 20 feet long in a fork-type mounting which allows the telescope to sweep all parts of the sky from the pole to as far south as declination minus 45 degrees. The combined weight of the fork and tube is over 12 tons. This whole assembly moves on two-inch ball bearings in the polar axis. The tube, partly cylindrical and partly conical and made of 5/16-inch welded steel plate, looks like a large mortar. The telescope shutter consists of two rotating shells located inside the tube behind the correcting plate. This construction allows the correcting plate to be removed or auxiliaries to be mounted without removing the shutters. The mirror and its cell are mounted at the lower end of the tube and are kept at a constant distance from the focal surface, regardless of temperature fluctuations, by means of three floating metal-alloy rods.

The telescope does not have positioning circles but instead employs Selsyn indicators which take their signals from declination and right-ascension gears and transmit the position electrically to the control desk. Other electrical features include automatic limit switches which stop the telescope four degrees from the horizon, automatic control of the dome's rotation and automatic regulation of the wind-screen height. The telescope is driven by a 1/25th-horsepower synchronous motor.

Two sizes of photographic plates are used in the camera: 10 inches square and 14 inches square. Since the plates must be bent to a spherical surface with a radius of 120 inches, to conform to the curved focal surface, they have to be very thin — less than one millimeter in thickness. The delicate glass plates are tested by bending beforehand to make sure they will not break in the telescope.

Construction of the telescope was finally completed in the autumn of 1948. The first tests showed that the telescope was much better than its specifications called for: the usable aperture of the correcting plate was actually 49.5 inches instead of 48. On 14-inch photographic plates the images were found to possess excellent definition over the entire field of 44 square degrees. The telescope is so fast that with 103a-O emulsion it reaches its limiting magnitude of 20.3 in about 12 minutes. This limit corresponds to the brightness of an average galaxy at a distance of 300 million light-years; in other words, the 48-inch Schmidt can "see" about one third as far as the 200-inch itself. So finally here was available a telescope of excellent quality that could photograph to great depths over a wide field.

THE SCHMIDT'S BIG PROJECT

The question now arose: What assignment should this great new instrument tackle first? There are three general types of job it is especially well qualified to do. One is to photograph wide, extended objects that an ordinary telescope can only sample piecemeal — such objects, for example, as the large galactic clouds of dark or luminescent gas called nebulosities. It was known that a few nebulosities covered several square degrees in the sky. Preliminary surveys with the new Schmidt

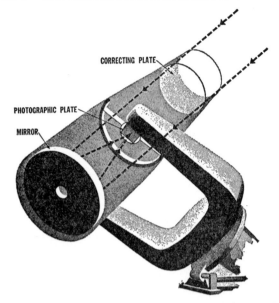

CORRECTING PLATE

PHOTOGRAPHIC PLATE

MIRROR

The Schmidt telescope uses a combination of lens and mirror. The images are focused on a photographic plate inside the tube. The Schmidt is a true camera, with no provision whatever for use with the human eye.

revealed that some of them were much more extensive than had been realized. Moreover, the new telescope disclosed new nebulosities so large that their identity would never have been suspected from the knothole views obtained with conventional telescopes. The Schmidt makes it possible to study the turbulence in these gas clouds as no other instrument could. Another type of large object that only the Schmidt can see anywhere nearly whole is a cluster of galaxies relatively close to us, such as the clusters in the constellations of Coma, Hydra and Virgo. It would take scores of plates with a reflector telescope to give the same coverage of these objects that four or five Schmidt plates afford.

The second kind of program for which the Schmidt is particularly suited is the statistical study of large numbers of objects. Statistical information about the distribution of stars in position, motion, brightness, color and so on is basic in the study of the structure of our galaxy. The Schmidt, used with various filter and emulsion combinations, can single out particular types of objects such as planetary nebulae and emission stars and chart their distribution. Similarly, for clues as to the structure of the universe the astronomer is interested in the distribution of systems outside our galaxy. This information will be made much more complete through the Schmidt.

The third appropriate big job for the Schmidt is a simple voyage of exploration. Its high speed combined with its wide field make this telescope ideal for patrolling the skies. Indeed, for this it is as superior to the conventional telescope as an airplane is to an automobile. The primary purpose of a patrol is discovery. The Big Schmidt can be used for two kinds of patrol. It can explore new regions out to fainter magnitudes than have been surveyed before; there it will undoubtedly discover many new faint objects. And it can rapidly resurvey parts of the sky already covered to detect changes there, *e.g.*, to discover new supernovae. The veteran 18-inch Schmidt on Palomar revealed 18 of these giant exploding stars in five years of patrolling certain galaxies.

When the Big Schmidt had proved its excellence in its preliminary tests, the Observatory's research committee, headed by Edwin P. Hubble, met to discuss the question of priority. It was decided that the best way to begin was to take the bull by the horns; namely, to undertake one extremely ambitious program — a systematic survey of the *entire sky* visible from Palomar. Such a project would at once cover many of the research programs awaiting the Schmidt. Not only would it provide an exploratory patrol of the skies but it would collect the observational material needed for the study of extended celestial objects and of distributions of stars and galaxies.

The National Geographic Society, which for over 60 years has sponsored expeditions to far corners of the earth in quest of geographic and scientific knowledge, became interested in this proposed exploration of the heavens. It therefore undertook the financial sponsorship of the survey, and the ambitious program was made possible as a cooperative under-taking of the Society and the Mount Wilson and Palomar Observatories. It was decided that photographic prints of each field should be distributed at cost to interested institutions and individuals. The whole set of prints will be known as the National Geographic Society Palomar Sky Atlas.

Hubble, the scientific director of the project, has pointed out that the Atlas will serve as a record of the heavens at one epoch, will provide an invaluable reference library for a great number of astronomical research projects, and, most important, will give us the first good look at the universe around us out to the distance to which the largest telescopes are working. Only one photographic atlas of the entire sky has ever been made. This was the Franklin Adams Survey conducted over 40 years ago with a small camera that reached only to the magnitude of 17.5. The Big Schmidt reaches out to stars 15 times fainter. In addition there is a tremendous gain in the improved definition and image quality of the Schmidt plates.

The survey with the Big Schmidt was formally inaugurated in July, 1949. The present objective is to photograph the sky to declination minus 24 degrees. It is hoped that in time funds can be obtained to construct a duplicate Schmidt in the Southern Hemisphere and complete the map for the entire sky.

The entire sky contains 41,259 square degrees. Of this, three-fourths is visible from Palomar. One 14-inch photographic plate used in the 48-inch Schmidt covers over 40 square degrees. This means that the Schmidt, even with fields overlapping, can cover the whole visible sky with less than 1,000 plates. Assuming that each field can be photographed in approximately one hour, it should take only about four years to map the sky.

THE SKY IN TWO COLORS

Each field is photographed twice — once with a blue-sensitive emulsion and once with a red-sensitive emulsion. The two exposures are made in rapid succession in order to obtain an accurate comparison of the sky in two colors. The two-color photography is in effect equivalent to a rough spectral analysis. Comparisons of starlight in the two colors provide what is known as the color index of the star. The color index provides a great deal of information about a star, since it indicates the star's spectral type and temperature. Colors of nebulosities are an aid in identifying the sources and mechanisms of their luminescence. Colors of extragalactic systems are helpful in identifying what kinds of stellar populations are present and aid in checking the remoteness of galaxies, for distant galaxies show considerable reddening because of the red shift. Colors also give a clue to the presence of obscuring material in interstellar space, because such material is more transparent to red light than to blue light. So in addition to the ordinary positional data available from a single photographic plate, the two color plates give appreciable physical data.

In its first few months the survey has already produced many interesting discoveries and accumulated much significant material. As expected, it has discovered new clusters of galaxies, new faint galaxies, new planetary nebulae, nebulosities, comets and asteroids. Some oddities and phenomena not yet explained have appeared. New information of possible cosmological significance has turned up, both on the distribution of extragalactic nebulae and on the density of matter in the universe. This material will be made available to astronomers everywhere in the near future. The process of systematic evaluation can then begin. It has been estimated that the Atlas as a whole will furnish so much information that astronomers will be kept busy for 50 years tabulating and interpreting it.

Great advances in scientific knowledge have been made either by the discovery of new objects or by looking at familiar objects in a new manner. The survey with the 48-inch Schmidt embodies both of these aspects. It may well prove to be one of the most significant astronomical endeavors of all time.

V
RESEARCH
AND REMINISCENCE

*T*he last of the seven great scientific world cruises of the non-magnetic ship Carnegie, operated by the Carnegie Institution, ended tragically on the afternoon of November 29, 1929. A first-hand narrative of the cruise by Dr. J. Harland Paul, ship's surgeon and observer, who participated in the voyage, is presented here.

The Carnegie's design, construction, and instruments were based on the experience of the Institution's Department of Terrestrial Magnetism with a chartered brigantine, the Galilee, operated from 1905 to 1908 during three cruises totalling 68,834 nautical miles in the Pacific Ocean. Launched in 1909, the Carnegie successfully completed, in six cruises, the first scientifically accurate magnetic-distribution survey and the first atmospheric-electric survey of the oceans between latitude 80° north and 60° south, traversing 252,702 nautical miles in 3,267 days at sea.

Of the 110,000 nautical miles planned for the seventh cruise, nearly half had been completed upon her arrival at Apia, Western Samoa, on November 28, 1929. The work had realized in practically every detail an extensive scientific program of observation in terrestrial magnetism, terrestrial electricity, chemical oceanography, physical oceanography, marine biology, and marine meteorology. Reports and records had been sent back to Carnegie Institution from ports along the way, so that in case of an accident no data would be lost. Captain J. P. Ault and his associates had good reason to be pleased and proud of their accomplishment.

Shortly after one o'clock on November 29th, while the ship was in the harbor at Apia completing the storage of 2,000 gallons of gasoline for the continuation of the cruise, an explosion took place, killing Captain Ault and a cabin boy, injuring five others, and destroying the Carnegie and all her equipment. The scientific personnel returned to Washington to prepare a comprehensive report on the data obtained, which was published by the Carnegie Institution in 1946.

Formerly of Marine Hospital, Ellis Island, New York, Dr. Paul was signed on March 20, 1928, for what turned out to be the Carnegie's last cruise. Then 28 years of age, and single, he had been educated at Columbia, the University of Chicago, Yale, and Harvard, having his bachelor's degree from Columbia and his M.D. from Harvard. His story of the cruise was included in the final Carnegie Institution report on the Carnegie.

280

J. Harland Paul

THE LAST CRUISE OF THE *CARNEGIE*

From *Results of Cruise VII of the* Carnegie; Carnegie Institution, 1946.

On May 1, 1928, the seventh cruise of the *Carnegie* began. Whistles roared from the harbor craft, and pleasure boats jockeyed for position to escort us down the Potomac. At midnight we reached the mouth of the St. Mary's River in Chesapeake Bay, and anchored till dawn. We were to spend four busy days here, "swinging ship," to be sure that our magnetic instruments and standard compass were not influenced by the new oceanographic equipment. A magnetic station had been set up on shore where simultaneous magnetic observations were made. To ensure ideal conditions for the land station, a magnetic survey of both sides of Chesapeake Bay had been completed a few days previously. Six "swings" of the ship on different headings were made, before everyone was satisfied that all was well.

The radio outfit was given its first trials here. Schedules were made with the Naval Research Laboratory and with headquarters of the American Radio Relay League. And throughout these four days, the atmospheric-electric instruments were being compared with similar ones ashore whose accuracy was well known. . . .

At dusk on May 5, all hands were summoned to heave up the anchor for the short trip to Hampton Roads — our first passage under sail. A stiff, steady breeze from astern bowled us along in grand style. Although we were not carrying full sail, we had the rare satisfaction of overtaking several steam vessels.

We were anchored off Newport News by eight o'clock next morning, and were greeted at once by "bum boats," little launches which were to be our inseparable companions in every port. They offered laundry service, taxis, provisions — everything we needed, and some things we did not. . . .

On May 10 we were towed out into the Roads, and set sails, while photographers on the tug made pictures. The breeze was just sufficient to give us steerage way. We had cast off our last ties with shore, and were at last headed for the open sea. Our last sight of land was Cape Henry at sunset.

It was a real relief to settle down to our ocean routine. The hectic past months gave place to as simple a life as possible. Meal hours were so arranged that in spite of their various duties, the staff could eat together. The radio operator and atmospheric-electric observers occasionally kept irregular schedules which made this not always possible. The watch officers and the engineer had their mess in the wardroom forward; and the forecastle was served from the same galley. The deck force was separated into two watches, as is usual on a sailing ship; the men spending four hours on and four off, with two "dogwatches" of two hours each between four and eight in the evening.

Our first morning out, May 11, was chosen for the first magnetic station. The ship was now fifty miles off the coast and away from local disturbances ashore. At sunrise the officer on watch calls the observers to the bridge for the declination observation. When they are assembled the ship's course is changed, if necessary, to keep the foresail from hiding the sun. Captain Ault and Torreson make readings of the marine collimating compass; Erickson measures altitudes of the sun with his sextant; and Scott enters each reading on special forms, with a time record for each observation. From these measurements we could tell how much the "variation" of the compass had changed since former cruises.

After breakfast is over, and when time sights on the sun have been made for longitude, the observers take their places at the magnetic instruments in the domes. Soule stands at the earth inductor; Torreson sits in the control room on the quarter-deck; and Paul reads aloud the heading of the ship from the standard compass in the chart room. This allows Soule to keep the rotating coil properly oriented. As Soule places the coil in various positions, Terreson reads the ammeter or potentiometer in the control room. From here he also starts and stops the constant-speed motor which rotates the coil. These observers determine the "dip" or inclination of the earth's magnetic field.

Meanwhile, Scott is in the after dome at the deflector. He places magnets of known strength near his compass and reads off their effect on it. Jones makes simultaneous readings of the standard compass in the chart room, and records for Scott. These two men measure the strength of the earth's magnetic field.

The afternoon is occupied in calculating the values for the magnetic elements. The observers were furnished special forms for recording, and these were so printed as to make the necessary tabulations as simple as possible. The formulae used in computing appeared in these, together with space for entering data derived from tables. By using these sheets it was practically impossible to overlook essential control records, such as air temperatures and chronometer readings. It is very easy to make these omissions when the observer's attention is directed primarily to the operation of the instrument itself.

For some of us the time-keeping on board was quite confusing at first. The ship's routine was operated on Local Apparent Time, with a resetting of clocks every morning at eleven. Many records were kept on Local Mean Time, others in Greenwich Mean Time. Then there was 75th Meridian Time for certain radio schedules, while a Sidereal-Time chronometer later became part of our equipment for gravity observations. In addition, for the most accurate time-signal comparisons, an "offset chronometer" was added, that loses one second in sixty-five of mean time.

After the evening time sight and the declination observation, we noticed a change in the color of the sea. It lost its grayish-green tint and became clear blue. The sea-water thermograph had shown great variations in temperature for several hours, and now read 75° Fahrenheit. At noon it had been only 46°. We were in the Gulf Stream. . . .

It was always a rule on the *Carnegie* to analyze and put in form the scientific data collected on each leg of the cruise, for the immediate use of hydrographers and oceanographic workers ashore. . . . For example, tables were drawn up showing the values of declination, horizontal intensity, and inclination, as given by the latest British, German, and American charts for the regions traversed by the ship. Against these we tabulated the measurements made on the voyage, so that errors in the charts might be corrected in future editions. Differences of as much as 1.°5 in declination were discovered on the passage from Newport News, with corresponding errors in the other elements. This serves to emphasize the importance of repeated surveys of the earth's magnetism, to determine the changes constantly taking place in the distribution of this mysterious natural force.

By early September our procedure at an oceanographic station had become somewhat standardized, and it might be of interest to describe just what takes place. On the morning of September 15, we are about two hundred miles from Barbados. At eight bells the new watch comes on deck and finds everything in readiness for heaving to. The winch is uncovered, the wires are threaded through blocks to the davits, outboard-platforms are in place, and running gear is laid out on deck ready for shortening sail. With the sound of the ship's bell still in our ears, the men dash to the tackle, blocks rattle and yards creak as the squaresails are taken in. The lower topsail alone is not furled, and is set aback to check our headway. Then one after another the fore-and-aft sails come down until only the mainsail and middle staysail remain. The ship is now hove to and comes up into the wind or falls off alternately with the helm alee.

The oceanographic team consists of four members of the scientific staff (Captain Ault, Soule, Seiwell, and Paul), the mate (Erickson), the engineer (Leyer), and the watch officer with his four seamen. Practically all operations take place on the quarter-deck. Mr. Erickson immediately attaches the bottom sampler to the piano wire, drops it over the stern, and signals to Leyer to pay out on the winch. Meanwhile Captain Ault and Soule are attaching the Nansen bottles, with their reversing thermometers to the aluminum-bronze wire. As these bottles are lowered one after the other in a long series, Paul reads the meter wheel. When the desired length of wire has been paid out he signals to Leyer to apply the brake. Another bottle is attached, more wire is paid out. This goes on till some eight or ten bottles are strung on at intervals of from five to five hundred meters.

At this station we are to reach down five thousand meters, so it will be necessary to send down two bottle series. The first, or "short series" will consist of nine bottles lowered to 5, 25, 50, 75, 100, 200, 300, 400, and 500 meters respectively, while one bottle is reversed at the surface. As the greatest difference in temperature and chemical salts occurs near the surface, the intervals are fairly short there. But in the "deep series," which is sent down later, the bottles are spaced 500 meters apart. The strain on the wire would be far too great were we to lower twenty bottles at once.

During this time Seiwell has put out the plankton nets. These are lowered in series, much as the bottles, but only three are used; one goes to 100 meters, another to 50 meters, and the third to the surface. Microscopic life in the sea is chiefly concentrated near the surface because sunlight does not penetrate water very far. All animals depend on plants for food, directly or indirectly, and of course it is sunlight which is utilized as a source of energy by plants such as diatoms.

Ten minutes are allowed for the lowered Nansen bottles to take up the temperature of their surroundings. Captain Ault now slides a brass "messenger" down the wire to reverse the first bottle in the series. As each bottle tips over, its own messenger is freed to proceed to the next bottle, and so on down the line. It takes from ten to forty minutes for the messenger to reach the lowest bottle. When they are inverted in this way, the valves automatically imprison a sample of water from the desired depth. Also, the mercury capillary of the thermometer separates in such a way that the temperature of that level can be read off on deck, no matter what temperatures are encountered on the way to the surface.

It is not possible to raise the bottle series until the bottom sampler has struck. With depths like five thousand meters this may take an hour. When the signal is given that the piano wire is slack, Leyer ceases to pay out, Erickson reads the meter wheel, and Captain Ault measures the vertical angle made by the wire. From these readings the depth can be calculated. Soule has meanwhile made an echo sounding to check this value.

The winch then brings up the bottle series and bottom snapper together. The bottles are removed from the wire and placed in sheltered racks. Paul collects water samples for chemical analysis, and Soule takes specimens for salinity determinations. When this is done, the deep-sea thermometers are read and the Nansen bottles prepared for their second plunge — this time to greater depths.

While all this is going on, Seiwell or Paul has put the plankton pump into operation. This apparatus is lowered three times, to levels corresponding to the depth of the tow nets. A measured volume of sea water passes through a fine silk net. The number of organisms captured, divided by the number of liters of water pumped, gives the "density of population" at each level. The plankton nets are hauled in after an hour or so. The specimens collected are preserved and labelled for future study.

It now remains to bring up the deep series and collect the sediment from the bottom sampler. This done, the sails are once more set and we proceed on our way. If everything has gone well there is still an hour before lunch in which to start the chemical work. The delicate hydrogen-ion tests are made first, to avoid the possibility of changes in the samples from contamination by the air or by sunlight. The other chemical characteristics are determined after lunch, along with the salinity.

These mornings are strenuous. There are many operations going on at once. Wires lead in all directions from the winch. The sun glares on the water, making it necessary to wear dark glasses. And only careful coordination saves us from utter confusion. Each man has his appointed tasks, but is always ready to lend a hand should things go wrong for the other fellow. And it was a rare day when something did not go awry. Wires might foul below the ship. Messengers might fail to reverse the bottles; or a "jellyfish" get in the way. The piano wire might snap, or the plankton pump fail to operate. Anything might happen, without warning, to upset the regular order.

In Barbados we found ideal conditions for trying out our diving helmet, and we made two expeditions to the reefs. For several of the men it was an entirely new experience. Only a poet could imagine the beauty and romance to be found under the waters of a coral reef. And certainly only a poet could describe what we saw in this fairyland of color and form. The dinghy is anchored at the selected spot, preferably in 15 to 30 feet of water, and the observer climbs over the side with a heavy copper helmet resting on his shoulders. A hose connected to a hand pump in the boat keeps him comfortably supplied with air, and he can wander about at will on the bottom.

One is in a new universe. Everything has a soft, ethereal outline except for the fishes that come to within an inch of the observer's nose to gaze at him in wonder through the plate-glass window. They are the most brilliantly colored of living creatures. One's sense of perspective seems to have been lost. Put out your hand to brace yourself on a coral head, and you find it far out of reach. Walking itself seems ridiculous; for in the water one's buoyancy is so great that the slightest spring upwards on the toes takes one off the bottom for a slow easy flight through space. Gravity has ceased to exist. Captain Ault described what he saw in a letter from which the following words are taken: ". . . schools of marvellously colored fish . . . forest of submarine trees waving in the water-surges . . . baskets of shell . . . jewel-cases of coral growth . . . grottoes of blue and sapphire . . . trees of growing coral with jewel tips . . . bristling, black-spined sea-urchins . . . a basket made of cocoa-nut-palm leaves gathered together at the top, perhaps full of treasure left by pirates . . . a wonder-world not reproduced elsewhere, not even in an aquarium."

Specimens were collected by the observers. A long screw driver and a heavy brass bucket were lowered on a rope, and on a signal from below the material was hauled up to the dinghy. Although the coral sand did not promise to be very rich in diatoms, we secured several bottles full for forwarding to Washington.

In the Pacific, after October 1928, the weather was perfect for pilot-balloon flights. The new equipment, supplied by the United States Navy, worked well and observations were made daily. With strong winds we were able to follow the balloon for only fifteen to twenty minutes, but sometimes it would be visible for an hour. By tying two together we could often follow them long after a single one would have been lost to view. In this way we traced the direction and force of the wind in the atmosphere up to heights of from two to six miles.

Three men take part in a balloon flight — usually Captain Ault, Torreson, and Scott. A pure rubber balloon is inflated with hydrogen from a tank, until it is about three feet in diameter. By "weighing" it we are able to calculate its rate of ascension. The scales operate upside down, of course, for the balloon pulls the pan upwards. At a signal from Scott, the recorder, the glistening globe is released. At one-minute intervals Torreson reads the azimuth, or horizontal position of the balloon with respect to the ship's heading; and Captain Ault checks the altitude by using an ordinary sextant. It was possible, of course, for Torreson to read off both altitude and azimuth from his theodolite; but the rolling of the ship often caused him to lose track of the object, while it was still clearly visible to the sextant observer. By reading the altitude from the sextant, it was possible for Torreson to sweep the sky at that level until he had again picked up the elusive sphere.

As a result of a multitude of observations on wind and weather conditions at sea, we have today fairly accurate "pilot charts" of the ocean, for the use of mariners. Now that transoceanic flying is coming to be a serious enterprise and not merely a stunt, it is highly important that aviators have "pilot charts" as well. They must know the direction and velocity of the wind at many levels, if they are to make successful flights over the great expanse of the ocean.

The month of February was a notable one for us in that we made several important changes in our instruments and methods. Ever since our departure from Washington, an attempt had been made to use the marine earth inductor for determining the strength of the earth's magnetic field in addition to the angle of inclination. All the trials up to the present time had failed to give results as reliable as those obtained with the standard "deflector." By changing the method slightly we now were getting comparable readings. . . .

The work with the pilot balloons was made very successful by the beautiful blue skies we enjoyed after clearing the dense clouds of the Peruvian coast. These flights often lasted thirty to sixty minutes, so one can imagine the severe strain on the muscles holding a heavy sextant for that length of time. It was necessary to devise some method for supporting the instrument. One of the deck chairs was fitted with arms and uprights to support an overhead bar. The instrument was suspended from this by a long, thin coil spring. In this way the entire weight was removed from the observer's arms, while still allowing freedom of motion. The whole outfit could easily be moved to whatever part of the deck was most favorable for observing the balloon. Captain Ault dubbed the device the "Joshua Chair," in honor of the Old Testament hero who commanded the sun to stand still. He had also suggested that it might better have been named in honor of Moses who at one critical moment in history had to call in the assistance of two men to support his arms.

Captain Ault says: "With this device we perhaps have carried the matter to an extreme, and caused the balloon to stand still. On at least three occasions, the balloon has suddenly appeared to be fixed in the sky, moving only very slowly in altitude and azimuth. On the first occasion, Torreson, the observer at the theodolite, was observing the balloon for fifteen minutes without getting much change. Finally Paul, who had been watching the flight, accused Captain Ault, the sextant man, of looking in the wrong direction and of reading altitudes that were far too low. It turned out that the theodolite had gotten sidetracked to Venus, and the difference between its altitudes of 76° and the altitudes by sextant of 45°, could no longer be ignored. On the second occasion both observers got sidetracked to Venus."

It is remarkable how closely a white balloon floating at a great height resembles the planet in the sunshine of the late morning or early afternoon. For most of us it was a great surprise to know that Venus could be seen at all in the middle of the day. Captain Ault told us that he had occasionally used this planet for determining geographical position at sea. This trick appears to have been known to mariners of former times, but has fallen out of use.

On February 8, Soule and Leyer moved the sonic depth finder from the radio laboratory to the control room on the quarter-deck. This was done to enable us to take additional night soundings without disturbing Jones who slept in the radio room. Paul had learned the technique of using the apparatus and now took a sounding after he had completed his Greenwich Mean Noon meteorological observations. Jones had by this time resumed a large number of schedules with amateur radio stations and had to get his sleep whenever he could, for he had regular magnetic observations and computations to do in the daytime.

New equipment was brought on board at San Francisco. Mr. Gish had tested out a new Kolhörster penetrating radiation apparatus in Pasadena and with Parkinson subjected it to further trials under the waters of Crystal Lake near San Francisco. This instrument registers the quantity of penetrating rays reaching the earth and may be lowered into the sea to determine the depth at which this powerful form of energy is absorbed. Mr. Gish also supervised the installation of a photographic conductivity recorder which had just been designed and constructed in our shop in Washington.

Forbush had brought with him several new chronometers and a photographic time-signal recorder with which time comparisons could be made accurately to one-tenth of a second and approximately to one-hundreth. These delicate time checks were necessary for the "gravity apparatus." He also brought new silk plankton nets for capturing organisms floating in the sea.

Graham had just come from the Scripps Institution in La Jolla where he had spent a month in studying the methods used in chemical oceanography. He and Dr. Moberg spent most of their time in San Francisco in reconditioning the oceanographic laboratory and in preparing new standard solutions. It was impossible to use the delicate chemical balance on board so these men set up the instrument on the pier. Graham also found time to calibrate the bottles which were to be used in determining the amount of oxygen in sea water. We had had such difficulty in obtaining distilled water of sufficient purity for our chemical work that it was decided to buy a small still of our own. Before Graham could take it on board he had to sign five copies of an affidavit that it would not be used for making liquor.

The gravity apparatus which was installed in the cabin by Dr. Wright was now to be tried out for the first time on a surface vessel. Cruises in Dutch and American submarines had shown that it might be expected to give reliable measurements if the roll of the ship did not exceed 10°. Besides this we were not bothered with constant vibration due to engines. The pendulum equipment was designed by Dr. Vening Meinesz of Holland and perhaps was the most delicate instrument on board. It recorded photographically the swings of three pendulums and recorded on the same paper the beats of a chronometer whose rate was known with great accuracy. From this trace the force of gravity at any place could be calculated. . . .

Forbush gave the gravity apparatus its first trials. As this instrument had never before been used on a surface vessel . . . difficulties were anticipated. They came — thick and fast. First, the heavy rolling threw a pendulum out of its support. On the next trial, it was found that the foot screws were not rigidly enough clamped down. Then it became apparent that some means must be devised for damping the motion of the apparatus. Finally, it was decided that only a new mounting would solve the difficulties. Notwithstanding these setbacks, several useful records were secured.

Heavy crosscurrents near the equator caused appalling losses of oceanographic equipment. On October 11 two silk nets were lost when the tow wire jumped its sheave and wore through. To avoid this trouble in the future, the rubber shock-absorber rope was attached directly at the forecastlehead, eliminating blocks entirely. The same day brought another accident, in which we lost a complete bottom-sampling and bottom-temperature outfit, through the catching of a splice in the meter wheel.

On October 19 we had to repeat the whole deep series of chemical and temperature determinations because a tiny piece of rope-yarn, caught by the messenger in descending, had prevented it from reversing the bottles. But on October 25 we were to suffer the most serious blow of all. The confusing currents below the surface entangled the bottom wire and the bottle series. In clearing them, the new aluminum-bronze cable was cut by catching on an outboard platform. We lost forty-two hundred meters of wire, nine reversing bottles, and eighteen of our precious deep-sea reversing thermometers. We could ill afford such depletions in equipment, so from this time on the thermal and chemical series was not lowered until the bottom sampling was completed. This change almost doubled the time required for a station.

After Graham joined the party, the chemical program was expanded to include determinations of silicates, phosphates, oxygen, and hydrogen ions at each station. With his help it was possible to add a vertical haul of a silk net from one hundred and fifty meters, at each station, besides occasionally checking the plankton pump. The pump determined the number of organisms floating in the water and to check its efficiency one filtered a known volume of sea water collected in a large bottle through a small silk net, and counted the marine plants and animals so captured.

On November 10, it was decided to heave to in the lee of Penrhyn Island to get a good measurement of the force of gravity. . . . This short stop enabled us to collect biological specimens and diatoms from the lagoon, and furnished a little recreation. This tiny atoll lies about midway between the Marquesas and Samoa, and is rarely visited by ships. The *Carnegie* had stopped there on a previous cruise, so that we were certain of a welcome from the white resident, Mr. Wilson. He was a castaway from the shipwrecked *Derby Park* in 1888, and since he has never left the island.

Once ashore we found, beside Mr. Wilson, a white merchant named Wilkinson, whom we had met in Tahiti in the spring; and a pearl trader by the name of Woonton. These men at once prepared a grand feast for us, while we rambled about the village, or fished the lagoon for specimens. Our hosts regaled us with many a South Sea yarn, as we sat on the verandahs drinking fresh coconut milk.

Two days later we made a similar call at Manihiki Island; here the gravity measurements were not so successful, owing to the swells coming in from the west. The Resident Agent, Mr. Williams, an old friend of a previous *Carnegie* cruise, gave us a hearty welcome to his charming island empire. This atoll offered a striking contrast to Penrhyn. Immaculate coral paths divided the neat little houses and flower gardens into "blocks." The natives were well dressed; the coconut palms were properly spaced and pruned for maximum production. Everywhere were evidences of a fatherly care on the part of old Mr. Williams. To the *Carnegie* this island is remembered chiefly for its characteristic dance. On a previous cruise photographs and moving pictures of this unique performance were destroyed by an accident in developing. And we were fated to lose ours for another reason.

We were now but a few days from Samoa, and the fast-dwindling supply of gasoline was eked out by catching every breath of air that blew our way. Reports and computations for the voyage about to close kept all hands at work till late at night.

The temperature of the ocean bottom had been measured at almost every oceanographic station since Honolulu, but just outside Samoa we recorded our lowest—one and one-tenth degrees centigrade. Another interesting observation was that in this region of long-continued calms, the surface may be almost a whole degree warmer than the water five meters below it; differences of one or two hundredths degrees are usual, when winds mix the surface layers. There was also a two-degree diurnal variation at the surface due to the sunshine.

The outstanding result of our echo sounding was the discovery of a new submarine ridge just north of Hawaii. We were able to show that there is no deep trough between Penrhyn and Manihiki, as the charts would lead one to believe.

The slopes of these two islands, as well as that of Tutuila, were carefully plotted.

Pilot-balloon flights had been very successful, thanks to the fine skies and the new theodolite. This instrument was so well adapted to conditions that the sextant chair designed by Captain Ault was seldom used.

Radio conditions had been unexcelled throughout the entire trip. Daily schedules with many amateurs in the United States, Hawaii, and Australia had brought us the news of the world, and had kept us in constant touch with our home office. . . .

Entering Pago Pago Harbor in the early afternoon of November 19, we did not have darkness to contend with as we did in the spring, when we nearly piled up on the reef. But this time the little engine was pushed to the limit in bucking the powerful wind squalls that swooped down from the mountains surrounding the bay. Time and again we were stopped dead in our tracks by these sudden gusts, almost losing steerageway at times. Because of the danger in tying up to the wharf under these conditions, we made fast to a buoy until the following morning.

The landing this time was almost a homecoming. Our friends of the spring were on hand to welcome us, with here and there a new face among them. The hospitality of the Naval Station was extended to us, as before. Since we were to remain here over a week, we had a better opportunity for observing Samoan life and for making collections on shore. Once the records and specimens were forwarded to headquarters, we found time to make several delightful excursions to native villages and into the mountains.

Graham and Paul spent the following Monday in collecting biological specimens. A guide was furnished by the chief who had entertained the party over the week end, and before they returned to the ship they had walked over a greater part of the island, crossing the mountains several times. A large number of native birds were secured for the National

Museum and a good collection of characteristic plants was made for the Carnegie Museum in Pittsburgh.

The day of our departure was drawing near and we had preparations to make. Supplies for the galleys and laboratories had to be stowed away and long-neglected letters answered. On November 27 we pushed off for Apia, arriving there on Thanksgiving morning, November 28.

On the morning of Friday, November 29, 1929, the *Carnegie* was at anchor in the harbor of Apia, Samoa. All morning Captain Ault and the remaining members of the staff were at work on board, the crew was engaged in loading the last of the barrels of gasoline into the ship's tanks. There remained only one hundred and fifty gallons to stow away when lunchtime came. After the noon meal, the crew resumed their task; Captain Ault unfolded a chair and sat on the quarter-deck; the engineer and mechanic were below in the engine room; and the others were scattered over the forward half of the ship, at various duties.

With a rumbling roar the ship was shaken from stem to stern by an explosion — then another. Captain Ault was thrown into the water. The men at work over the tank room were hurled to different parts of the ship. The engineer and mechanic were trapped in the engine room and in a moment the whole quarter-deck was enveloped in flame.

The steward and Soule, rushing on deck, dived overboard to save the Captain. The engineer and mechanic fought their way out of the blazing engine room by raising themselves through the gaping hole in the deck. The uninjured men dragged the others free of the flames. To save the vessel was out of the question and all attention was directed to the saving of lives.

Small boats had been launched at once from the other ships in the harbor. Captain Ault, who had been holding on to a rope as he floated in the water, was helped into one of these and with the other injured men was taken ashore. Apparently he was suffering only minor injuries; but his injuries were serious and on the way to the hospital, our Captain died as a result of them and of shock.

The other men who had been on the quarter-deck suffered fractures and severe burns. They were given immediate surgical attention by the hospital staff, who had been notified by telephone of the accident.

When the survivors were collected ashore, Tony, the cabin boy, could not be accounted for. He had last been seen in the after galley, immediately next to the tank room; so it was apparent that he too had lost his life. His remains were not discovered until December 4, when salvage operations on the charred hull of the vessel were commenced.

Seaton, Graham, and Paul had been away on a collecting trip and did not return until about three hours after the tragedy. The hospital staff and Government officials had done everything in their power for the survivors. There was nothing further to do but to await the arrival of the U.S.S. *Ontario,* the naval vessel from Pago Pago which the Navy had ordered to our aid.

The engineer and mechanic were too severely burned to stand the journey to Pago Pago, so they were left in the hospital at Apia. Parkinson, as second in command, also stayed to take charge of affairs there. On the day following the explosion, all the others were taken to American Samoa to await the steamer from Sydney. The three injured seamen we brought with us were put in the Naval hospital while the members of the staff were taken into the homes of the Naval officers, and the crew was quartered in barracks.

Everything was done to make us comfortable. We were furnished necessary clothing — for the ship and all its equipment together with our personal effects, had been a total loss. Governor Lincoln, on behalf of the Navy, arranged immigration papers for entry into the United States for those who were not citizens.

On December 6, the survivors accompanied the body of Captain Ault aboard the *Ventura* for the sad journey home.

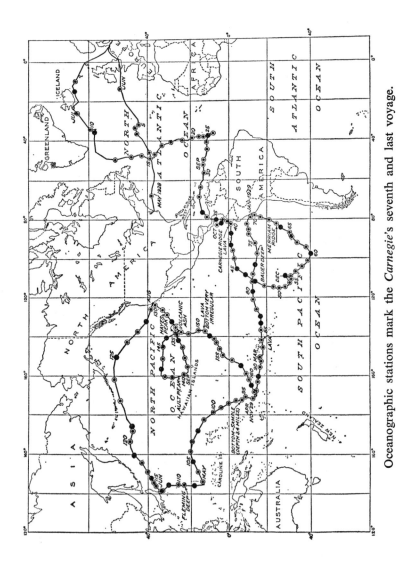

Oceanographic stations mark the *Carnegie*'s seventh and last voyage.

A most unusual man was Dr. Walter Sydney Adams, astronomer, whose absorbing account of early days at Mount Wilson
*Observatory, somewhat abridged, is included in these selections.
His reminiscences were written originally for the amusement and
information of the staff of the Observatory, but attracted so much
favorable comment they were later published in the October and
December 1947 issues of* Publications of the Astronomical Society
of the Pacific.

*Dr. Adams, the son of New England missionaries, was born
in 1876 in Kessab, North Syria, a small town on the slopes of
Mount Cassius near the ancient city of Antioch. Although he left
Syria at the age of eight, he maintained an interest in the East and
in classical languages throughout his life.*

*He received his A.B. degree from Dartmouth College in
1898, and his A.M. degree from the University of Chicago in 1900.
He continued his studies at the University of Munich, Germany,
until 1901, in that year joining the staff of Yerkes Observatory, at
the invitation of Dr. George Ellery Hale. He remained at Yerkes
until the spring of 1904, when Dr. Hale, who was then forming
the staff of the Mount Wilson Observatory, asked Adams to move
west with him and become a member of the pioneer group at
the new Observatory, which also included G. W. Ritchey and
F. Ellerman.*

*During the following 42 years he was a member of the
Mount Wilson staff, and from 1923 until 1946 he was its Director.
He took never-ending delight in the primitive life atop Mount Wilson in the early days of the Observatory, pleasure in roughing it
being a heritage from his childhood in Syria and later in New
Hampshire. He loved the mountains and gladly risked the tortures
of poison oak and the hazards of climbing to ascend the peaks
of Mount San Antonio and Mount Whitney. At his home he kept
a collection of rocks to remind him of places where he had been.*

*Both at Yerkes and at Mount Wilson he conducted ex-
tensive and productive investigations into stellar velocity and luminosity with the aid of the spectrograph, and he had a large part in
the design of the 100-inch and 200-inch telescopes. In the course
of his career he published some 300 technical papers on various
aspects of astronomy. He continued his research for ten years
after his retirement in 1946, and died at his home in Pasadena,
California, in 1956.*

Walter S. Adams

EARLY DAYS AT MOUNT WILSON

From *Publications of the Astronomical Society of the Pacific,*
October and December 1947.

The history of the origin and development of the
Mount Wilson Observatory is primarily the story of the insight, enthusiasm, and courage of a single individual. Those
who were fortunate enough to be associated with George Hale
at the Yerkes Observatory, and to come with him to California when the Mount Wilson project began to develop, had
long recognized these qualities as well as the rare personal
charm and highly cultured intelligence which made him such
a delightful companion on every occasion. But even they were
hardly prepared for his reaction to the novel and primitive
conditions which were encountered during the early years on
Mount Wilson. Apparently combined with a deep-seated love
of nature in every form was the spirit of the pioneer, whose
greatest joy is the adventure of starting with little and taking
an active personal part in every phase of creation and growth.
To both of these inborn characteristics of Hale, Mount Wilson
in 1904 offered a rich field and full scope for their employment.

It is difficult for anyone in later years to realize how
simple were conditions in southern California when the first
steps were taken in the establishment of the Mount Wilson
Observatory. Pasadena was a town of about fifteen thousand
inhabitants spread widely over a large area. Open country
with vineyards and orange groves intersected by rambling
dirt roads occupied what is now the closely built area along
North Lake Avenue, while eastward of this street were only

301

occasional farmhouses and small dwellings set in a semidesert environment. Small private water companies furnished a more or less intermittent service to their subscribers, and water shortages followed almost invariably upon winters of deficient rainfall. A prominent and rather amusing feature of what might be called the business district of Pasadena, then essentially limited to Colorado Street between Fair Oaks and Marengo Avenue, was the elevated wooden structure beginning at Colorado Street and extending southward as far as Raymond Hill. This was a portion of a proposed cycleway by means of which residents of Pasadena could, for a moderate toll charge, ride their bicycles into Los Angeles over a wooden runway free from the difficulties of roads deep in sand and the inconveniences of horse-drawn traffic. The enterprising company which conceived this project was so unfortunate as to start operations just at the beginning of the motorcar age, but may be said to have anticipated to some degree the modern Parkway.

As might be expected, access to Mount Wilson and conditions on the mountain top were equally simple and verged upon the primitive. Two trails led to the summit: the first, the old Indian trail leading from Sierra Madre up the canyon of the Little Santa Anita stream; and the second, the trail built by the Pasadena and Mount Wilson Toll Road Company, beginning at the mouth of Eaton's Canyon in Altadena and zigzagging up the rugged southern face of the mountain range. The ambitious name of "toll road" was given to this trail, averaging about two feet in width, but it was not until it was widened by the Observatory in 1907 to provide for the transportation of the 60-inch telescope that it became in any sense an actual road. The two trails were for many years familiarly known as the "old" and the "new" trails, respectively.

On the summit of the mountain the only building was the so-called "casino," a log cabin, built of cedar logs in 1893 and abandoned after a few years. At the time of Hale's first visit to the mountain in 1903 the roof had partially fallen in, and he often spoke of the first night he spent within the building lying on a cot and watching the stars crossing a hole in the roof some twenty feet in length. The heavy log walls, however, were in good condition, and it was clear that simple repairs would make the cabin at least temporarily habitable for a small group. . . .

Transportation in the mountains at this time was, of course, wholly by pack train, and the general term of "animal" was applied to the miscellaneous and picturesque assortment of burros, mules, and occasional horses which were maintained in a stable or corral at the foot of the old trail in Sierra Madre for carrying visitors and supplies to the various mountain camps. "Ordering an animal" was the regular expression for engaging a beast of burden in case the visitor did not wish to face the rigors of the eight-mile climb to the summit of Mount Wilson wholly on foot. But as experience often showed, the passenger on a mule or burro fully earned his passage. Books could be written about the personal characteristics of these sagacious beasts and the infinite variety in their individual behavior. One would deliberately expand his chest when the saddle was placed upon him so that the rider, after a good start, would presently find the saddle rolling beneath him at some awkward point in the trail; another would groan heavily when the grade became steep, but if the rider once dismounted, he would be fortunate to overtake his mule within several miles; while still a third would show an almost irresistible desire to roll over, frequently selecting a stream bed for this purpose.

The names of these animals had in most cases been bestowed upon them by the pack-train drivers, a somewhat philosophical group of individuals skillful in throwing a diamond hitch, extremely resourceful, but owing to their profession endowed with a lurid vocabulary and a distinctly pessimistic outlook upon life. They had to be prepared for the worst in dealing with their charges and usually encountered it. Nevertheless, the drivers were proud of their animals, learned their characteristics and various abilities, and took excellent care of them in times of trouble. Each problem of transportation was solved separately, one mule being selected for his riding qualities, a second for his strength, and a third for his ability to carry a top pack or to swing widely at the turns of the trail. An astonishing feat of skillful planning was the successful transport of a 225-pound metal bathtub to the halfway house on the old trail. For several years one particular mule made a regular practice of carrying up this trail a resident of the Mount Wilson Hotel whose average weight was 268 pounds. The burros that carried lighter loads handled most of the material dealt with in quantity, such as cement, lime, lumber, and structural steel. The limits set by such methods of transportation necessarily affected every feature of the design of the earlier buildings of the Observatory, and it is of interest to realize that no single structural member of the Snow telescope building exceeds eight feet in length. Similarly, the wooden doors of the original Monastery, which was burned in December 1909, were limited to a width which would permit of their transportation on the sides of a burro without dragging on the ground. As it was, several of these doors when installed were found to be slightly rounded at the corners where they had touched the ground occasionally during the long journey up the trail. . . .

The story of the foundation of the Observatory is so well known that it requires little comment here. It passed through three phases which followed closely upon one another. At first it was simply an exploratory expedition to investigate observing conditions on Mount Wilson, and was financed personally by Hale and a few individual gifts. At this time a small coelostat was erected near the casino on the present hotel site, and photographic observations of the sun were carried on directly as well as spectroscopically with a long instrument placed in a temporary tube north of the coelostat. They were supplemented by visual observations with a portable four-inch refracting telescope equipped with a polarizing eyepiece. The results were so satisfactory that an application was made by Hale to the Carnegie Institution for a grant to bring the Snow telescope of the Yerkes Observatory to Mount Wilson on an expeditionary basis. The grant, amounting to $10,000, was made in April 1904, and most of the summer of that year was devoted to laying the foundations, constructing the piers, and designing the building for this instrument. Before its completion, however, the confidence which President Woodward and the Trustees of the Institution placed in Hale and his plans, together with a change from their previous policy of making a large number of small grants to investigators throughout the country to one of establishing major departments within the Institution itself, resulted in the appropriation of a large amount for the establishment of an independent observatory. The third and final step had been taken, and the Solar Observatory, as it was known during the years preceding the completion and operation of the 60-inch telescope, began its interesting career. The hopes of its courageous founder had been richly fulfilled.

The members of the staff of the Yerkes Observatory who came to Mount Wilson with the Snow telescope and remained to take part in the establishment of the new observatory were G. W. Ritchey, F. Ellerman, and Adams. Francis Pease came about one year later. Ritchey had been instructor at a Chicago manual-training school, and was primarily a draftsman and designer. At the Yerkes Observatory he developed much skill in optical work and also became an able observer, especially with the 24-inch reflector in the design and operation of which he took a leading part. At Mount Wilson he designed much of the earlier equipment, more especially the 60-inch telescope and dome which have now been in active operation for more than forty years with comparatively little modification. He also carried through successfully the figuring of the 60-inch and 100-inch mirrors together with their auxiliaries. Ritchey's main observing interest lay in direct photography, and his absorption in this field was so intense that it handicapped to some extent the design of instruments which had to be used for many different purposes. As an observer he was most painstaking, and the admirable photographs which he obtained in the early years of the 60-inch telescope have proved of great value in later investigations. In personal relations Ritchey was somewhat difficult, and it is a tribute to Hale's unfailing tact and consideration that their association lasted throughout so many years.

Ellerman was of quite a different type. His long association with Hale at the Kenwood and Yerkes Observatories, his observing and photographic ability and experience, together with his ingenuity and mechanical skill, made him invaluable, especially during the years of construction and development on Mount Wilson. If anything went wrong with an instrument he could almost invariably repair it or at least make it operate temporarily. This ability was of immense value at a time when transportation was slow and difficult, and the fifteen miles separating the Observatory from its instrument shop in Pasadena formed a gap which it took the better part of a day to cross. So during these early years we depended upon Ellerman in a thousand ways, and it was a rare occasion when he could not meet the emergency.

As was the case with the other members of the small Yerkes group, the somewhat wild and primitive conditions on Mount Wilson were quite new to Ellerman and he enjoyed them greatly. He had seen but few mountains previously, and the views with their outlook upon the valley and the deep canyons, the new trees and flowers, and occasional glimpses of wild animals all made a strong appeal to his love of nature. He would have ranked high on John Buchan's criterion that "there is something wrong with the man who sees a high mountain and does not want to climb it," for Ellerman was an ardent mountain climber. The occasional rattlesnake and the almost mythical mountain lion provided the element of excitement, and he made elaborate preparations for meeting them. On my first trip up Mount Wilson, Ellerman and I met at the foot of the new trail. He wore a "ten-gallon hat," high mountain boots, and a full cartridge belt from which hung a revolver on one side and a hunting knife on the other. Naturally I was greatly impressed and pictured a struggle for existence on the wild mountain top, which bore little resemblance to later actuality. Together we explored many trails and climbed the higher mountains to the eastward. Ellerman was a most pleasant companion on all such expeditions, and recollections of these walks and of occasional games of golf in the valley form some of the pleasantest memories of these early years on the mountain.

Pease was an able designer and optician and developed greatly as his responsibilities grew. His patience knew no limit, a quality which stood him in good stead during the earlier years of his association with the Observatory. He became a remarkably skillful observer, devoting to his work exceptional care and thought and all the resources of his technical knowledge. His affection for the Observatory was profound, and he spared neither time nor strength in its service. As a comrade and friend he was unfailingly cheerful at all times, and his sudden death in later years brought to everyone a feeling of personal loss and full realization of his notable contributions to the success of the Observatory.

The heart and center of the small group was George Hale, then in the prime of life and full of enthusiasm over the almost boundless prospects opened out by the generous action of the Carnegie Institution. In the establishment of the Observatory he found the complete fulfillment of his ideal of an institution devoted purely to research, free from many of the restrictions imposed by university affiliations, and able to build its equipment to fit the problems he had in mind instead of seeking the problems to fit existing or preconceived instruments. In addition, the mountain in all its varying aspects was a source of perpetual delight to him. Its quiet and peace, the sense of remoteness and isolation, the changing views, and the brilliant skies by day and night were a constant joy. He often called himself a sun-worshiper, and nothing pleased him more than to leave Pasadena on a summer morning when the valley was covered by fog, and halfway up the trail to burst out into the bright sunshine and see the distant mountain peaks outlined against the deep blue sky. He enjoyed, too, all the simple activities of these early days, the exploration of the mountain, examination of the water supply, the selection of sites for buildings, and the erection and testing of the instruments, most of them temporary in character, which were used for much of the earlier work. From his boyhood he had loved to work with machine tools and had developed very considerable skill in the use of the lathe and milling machine. This skill he put to active use on the mountain, and many a temporary slit or lens or grating mounting of this period was built by Hale with his own hands.

The winter of 1903–1904 was spent by Hale and his family in Pasadena. He then returned to the Yerkes Observatory for a brief stay and, on coming back early in March, brought Ellerman with him to begin systematic observations on Mount Wilson with the small coelostat. Late in April, Hale went again to the Yerkes Observatory and this time on his return to Pasadena brought with him Ritchey and Adams. The grant from the Carnegie Institution for the temporary transfer of the Snow telescope to Mount Wilson had just been made and the first steps in the establishment of the new observatory may be said to date from that time. The railway journey from Chicago to Pasadena required four nights and three days, with forty-minute stops at suitable intervals at dining stations along the way. Hale's interest and excitement increased as we neared the end of the journey, for he realized that the decision had been made, and that a new life with new responsibilities and opportunities lay before us all. I remember that the three of us climbed out of our berths at dawn as the train was going through the Cajon Pass, in order to enjoy the abrupt and always spectacular transition from sheer desert to the gardens and orange groves of the coastal plain of southern California.

The plan of organization at the outset was for Ritchey to be in charge of a small machine shop in Pasadena to build auxiliary instruments for the Snow telescope, for Ellerman and Adams to continue the solar observations on Mount Wilson and to assist in the design and erection of the building to house the telescope, and for Hale to divide his time between Pasadena and the mountain. . . .

On Mount Wilson the only existing building was the log cabin called for some unknown reason the "casino." It had been made comfortably habitable by the time Ellerman arrived on the mountain, through patching the holes in the roof, repairing the floors, and dividing a large room into small cubicles with building-paper partitions. The outstanding feature of the casino was the huge fireplace in the living room built of large granite blocks and capable of holding logs two feet in diameter. The fireplace and chimney were the handiwork of George Jones, a stonemason by trade, whom Hale had discovered in Pasadena. It soon appeared that Jones could turn his hand to every type of construction and development required by the Observatory, and as superintendent in later years he was responsible for the widening of the new trail into a serviceable road, the transportation of all materials to the mountain, and the erection of every important building and instrument, including the dome and the mounting of the 100-inch telescope. The debt which the Observatory owes to the genial, resourceful, and efficient George Jones is beyond estimate.

The group living in the casino between May and December, 1904, consisted of Hale, during his frequent visits, Ellerman, Adams, and the cook, Wilbur by name. . . . Life was very pleasant on the mountain during this first summer. A regular series of meteorological records was kept, and observations were made of the solar seeing throughout the day. The small coelostat was used for direct photography of the sun and to provide an image on the slit of a temporary spectrograph, the focal length of which could be varied up to as much as sixty feet. The purpose of these latter observations was to test the effect of seeing within the spectrograph itself, and thus to provide a basis for the design of the permanent instruments to be used with the Snow telescope. Early in the summer the site for this telescope had been selected, and soon operations were begun on leveling the ground, laying the foundations, and building the large south pier. Meanwhile, explorations of the top of the mountain for the location of other instruments were in progress. Although the Institution had not yet acted upon the larger project involving the completion of the 60-inch telescope, the prospects appeared favorable, and the selection of a site for this instrument was considered. For a time the choice lay between its present site and the prominent knoll at the northeast corner of the mountain top. Accessibility and less proximity to steep mountain slopes were the determining factors in the selection of the present site.

It was at this time, too, that Hale and Adams set out one day to explore the ridge extending southward from the site of the Snow telescope. Small hatchets were taken along to cut a path through the brush since no trail existed. After working our way slowly downward for a quarter of a mile, we came out upon a small opening free from brush at the very end of the ridge. The ground was nearly level, with several large spruce trees and a little grass in the cracks of the exposed ridges. On three sides the slopes fell abruptly into nearly sheer precipices, and the view of the valley, the canyons, and the distant mountains was magnificent and quite unobstructed. Hale's enthusiasm was tremendous. "This is where we must have the Monastery," he said; and six months later on a warm and quiet December evening we moved into the attractive building around which so much of our life was to center for several years to come.

While living at the casino we were almost in the path of the visitors who were hardy enough to undertake the eight-mile trip up the old trail from Sierra Madre. About ten days after our arrival, a district convention of the Order of Elks was held at Strain's Camp, about half a mile distant on the north side of the mountain. The officials who planned this outing had clearly made a mistake. The comfortable physique of the average Elk is not at all adapted to several hours of severe mountain climbing, but since the supply of burros and mules at the foot of the trail could provide for only a fraction of the group, no choice remained for the others but to attempt the journey on foot. As a result, we at the casino had to operate something in the nature of a first-aid station for exhausted Elks during much of one afternoon and evening. Some were still staggering along on their feet while others lay on the ground vowing they could go no farther even if their lives depended on it. The last arrival had been nine hours upon the trail, a record which probably remains good to this day. . . .

Wild animals, with the single exception of deer, were more plentiful on Mount Wilson in these early years than at present, partly because relatively few visitors came to the mountain, and partly because natural conditions were undisturbed. There was a tradition that even an occasional bear was to be seen in the vicinity, although we never encountered one. Hunting was allowed at all times, a fact which doubtless accounted for the comparative absence of deer, and a simple request was sufficient to obtain a permit to carry a revolver or other firearms. As a result we had a small arsenal at the casino and often practiced target shooting, at which Ellerman became especially proficient. The stump of an old pine not far from the casino should some day prove to be a lead mine. Adventures with rattlesnakes were fairly common, and once we had to rescue the casino cat from a coiled rattlesnake he was starting to investigate.

One of our problems at this time was the telephone line which was a constant source of uncertainty, consisting as it did of a single iron wire stretched along on bushes and occasional trees. Even when it was in normal working condition the operator had to shout so loudly that it sometimes seemed doubtful whether the listener did not hear the voice directly rather than through the instrument. We had many exasperating experiences with it and spent considerable time in trying to obtain a satisfactory ground connection, no easy task in the dry decomposed granite of the mountain top. One day soon after we had moved into the Monastery a snowstorm crippled the telephone line and Hale and I started out to try to locate the trouble. We found a broken wire on an oak tree on the ridge near the site of the present 10-inch telescope, and Hale at once started up the tree in the driving sleet to repair the break. As he worked, his enthusiasm, which neither weather nor numb fingers could affect in the least, broke out in the words of his favorite poet, and I heard, "It was a storm from fairyland . . . ." coming down from the tree above the roar of the wind. . . .

The construction of the Snow telescope building and the erection of the instrument were the chief activities of the summer and autumn of 1904. As has been mentioned already, transportation by pack train set a definite limit of eight feet to the length of all structural-steel members in the building. The only exception was the steel bar, ten feet long, to which the windlass for moving the small sliding house over the coelostat was attached, and this required extensive negotiations and the use of a top pack on a selected animal. The transportation of the heavier parts of the telescope mounting, however, presented a difficult problem, since some of them were single pieces weighing as much as 350 pounds and quite beyond the capacity of even the strongest mule. This afforded an interesting opportunity for the exercise of Hale's ingenuity. He designed a small steel truck about ten feet long, with a width and tread of twenty inches. Small wheels with rubber tires were used and the body was underslung to keep the load close to the ground. The truck was steered from both ends so that it could negotiate the sharp turns in the trail. The motive power was a mule attached to the front end. In operation the truck was a remarkable sight. First came a man leading a mule, next the truck and its load with a man at either end to steer the vehicle, and finally a second mule led in the rear to act as a reserve, or, in case of need, to pull the truck back upon the trail. This unique invention proved most successful and made numerous trips up the new trail, bringing, in addition to the mirrors and mounting of the Snow telescope, the heavy flywheel of the first gas engine to provide electric power on the mountain, and parts of several machine tools. Naturally the truck was slow in operation and a round trip from the foot of the trail required a full day.

Designing the Snow telescope building to insure the best use of the instrument was an important problem, and many experimental tests were made before a final decision was reached. Two related factors were involved. The first had to do with the heat radiation from the ground as affecting the quality of the solar image; the second with the protection of the beam of light inside the house during its long passage from the coelostat to the image-forming concave mirror and back to the focus. It was clear that soon after sunrise the warming of the ground, especially where there was little coverage by trees or brush, would set up convection currents of warm air which were certain to injure the definition of the image. The obvious solution, of course, was to place the telescope and especially the coelostat high above the ground, but considerations of expense and the fact that the Snow telescope was planned to operate in at least an approximately horizontal position set a limit upon the height which could be attained. Some compromise was evidently necessary. So tests were made with the four-inch portable telescope of the quality of the image at various heights above the ground, platforms built on temporary wooden towers being used for the purpose. Hale even climbed a large pine tree near the small reservoir, dragging the observing telescope with a rope to a height of some sixty feet. The ground coverage was also examined and the effect of shielding bare ground in the vicinity of the coelostat with cheesecloth to reduce heating and radiation was tested repeatedly. The final decision was to place the building on the highest point of the selected site, with the coelostat on a stone pier about twenty-five feet in height on the slope below the high point. The height of the coelostat was such that the beam of light to the concave mirror as well as the building itself sloped downward to the north by a very few degrees from the horizontal, and conformed to the contour of the ridge.

The necessity for shielding the beam of sunlight within the building from injurious convection currents led Hale to adopt the louver type of construction for the walls. The successive skirt-like louvers prevented direct sunlight from reaching the interior of the building but permitted free circulation of air around the closed inner wall. A series of ventilators in the roof allowed the warm air to escape upward.

It is interesting to note how directly Hale's experience with the Snow telescope led him to the design of the vertical tower telescope two or three years later. The tower provided a simple means for obtaining the incident sunlight at a high elevation relatively free from the convection currents near the ground; the lens gave a convergent vertical beam much less susceptible to disturbances than one parallel to the ground; and the vertical spectrograph offered great convenience in use, excellent conditions of temperature control, and the possibility of using much greater focal lengths than are possible for horizontal spectrographs except under the best laboratory conditions.

Since these casual reminiscences are in no sense a history of the scientific work of the Observatory, I am making no attempt to describe in detail the uses to which the Snow telescope was put after completion. Undoubtedly the single most fruitful investigation associated with the instrument was the study of the spectrum of sunspots. To the best of my knowledge the first photograph of a spot spectrum ever obtained was made with this telescope in 1905. A series of excellent spectroheliograms led to the detection of solar vortices, and these in turn to Hale's dramatic discovery with the tower telescope of the magnetic field in sunspots. It is perhaps of interest to note that the first application of high dispersion to stellar spectra was with the Snow telescope. A plane grating, cemented into a metal box filled with water which was maintained at constant temperature, was used with a 13-foot spectrograph on two or three of the brighter stars. Since the grating was not exceptionally bright and the telescope mirrors were usually somewhat tarnished, exposure times were extremely long. A very respectable spectrogram of Arcturus, taken with a total exposure of twenty-four hours on four consecutive nights, is still in existence, however. A comparison with the twenty-minute exposures of Arcturus made with the 100-inch telescope and its coudé spectrograph, which gives considerably larger scale and finer definition, affords a slight illustration of the progress of the Observatory since those distant years. . . .

During the first year of our occupancy of the Monastery, Professor Barnard came out from the Yerkes Observatory on a special expedition, bringing with him the Bruce telescope to photograph a portion of the southern Milky Way. Of course all of us knew him well and it was a delight to have him with us on the mountain. He at once fell in love with the mountain and everything connected with it: he was fascinated by the views, studied the birds, measured the growth of yucca stalks, and treasured the sight of a deer. I remember his excitement one winter morning when he came in to breakfast and announced that he had just seen a wildcat walking through the snow outside his bedroom window. . . . Barnard's devotion to the mountain may be judged by the fact that during four months of his stay he made but one trip to the valley. This was to Sierra Madre to see a notary and to have his hair cut, after which he turned around and started back up the trail. . . .

Barnard's hours of work would have horrified any medical man. Sleep he considered a sheer waste of time, and for long intervals would forget it altogether. After observing until midnight, he would drink a large quantity of coffee, work the remainder of the night, develop his photographs, and then join the solar observers at breakfast. The morning he would spend in washing his plates, which was done by successive changes of water, since running water was not yet available. On rare occasions he would take a nap in the afternoon, but usually he would spend the time around his telescope. He liked to sing, although far from gifted in the art, but reserved his singing for times when he was feeling particularly cheerful. Accordingly, when we at the Monastery heard various doleful sounds coming down the slope from the direction of the Bruce telescope, we knew that everything was going well and that the seeing was good.

The Bruce telescope was housed in a small wooden building with a floor about three feet above the ground. A trap door in this floor gave access to the weights of the driving clock which were suspended by a cable. When the telescope was pointed high in the sky, Barnard often found it convenient when guiding to open the trap door and sit on the floor with his legs dangling through the opening. One summer morning at breakfast he casually mentioned to us that he had heard some odd rustling sounds beneath him during several preceding nights. On examination we found and succeeded in killing a fair-sized rattlesnake which had been making his home below the floor. Whether the snake had ever attempted to investigate the intruding legs we never knew, but Barnard took the episode quite calmly and on the following night the trap door was open as usual. . . .

In 1905, Henry Gale, then instructor in physics at the University of Chicago, was invited by Hale to come to Mount Wilson and spend a year at the small spectroscopic laboratory which had just been completed on the mountain. This was the beginning of a friendship which extended over many years. The first photographs of sunspot spectra had just been obtained and we worked together in the laboratory in the attempt to duplicate the differences among groups of lines observed in spots. The sources used included the direct-current arc at various current strengths, the rotating arc recently invented by

Henry Crew, and finally the core and the outer flame of the arc. The experiments proved successful and led to the classification of lines into groups according to temperature, which has had many applications in stellar spectroscopy and in the analysis of the atomic spectra of the chemical elements. Gale's powerful figure surmounted by a huge curved pipe was one of the familiar and most pleasant sights of that second winter on the mountain, the first at the Monastery. At this time the Mount Wilson Hotel was visited by us frequently, usually for the purchase of tobacco. The short, steep climb toward the laboratory on the return journey worried Gale, and at his suggestion the two of us undertook the construction of a trail of nearly uniform grade, beginning near the battery room and extending across the slope below the laboratory and around a sharp point to the dividing line with the hotel grounds. Gale worked manfully on this project, which when completed was named the "Lucky Strike Trail" after the brand of tobacco usually found at the journey's end. The trail was used for several years but in the course of time was almost completely obliterated by winter storms.

The group at the Monastery had many good times together during this period. Abbot* and his assistant, Ingersoll, of the physics department of the University of Wisconsin, were making the first of the many Smithsonian expeditions to Mount Wilson, and their presence added greatly to the pleasure of life on the mountain. Hale had introduced Abbot to the Oriental stories in Beckford's book on the monasteries of the Levant, and our evenings usually started off with a dramatic rendering by Abbot of the tale of the Jew of Constantinople and Solomon's Seal which he knew by heart. Occasionally the Smithsonian challenged all comers to a game of duplicate whist, but more often the group would gather around the fireplace for discussions of plans of work or of the state of the

---

* Dr. Charles Greeley Abbot, at this time Director of the Smithsonian Astrophysical Observatory, later Secretary of the Smithsonian Institution.

world in general. Hale's amazing breadth of interests, his great personal charm, and his stories of important figures in science and national and international affairs make these evenings stand out in memory. . . .

Of the many pleasant recollections of Abbot's visits to the mountain, one deserves mention as an illustration both of his sense of humor and of his excellent upbringing as a youth. Transportation to Mount Wilson had passed through the successive stages of walking, riding a burro or mule, or of traveling in a mule-drawn wagon. Finally, in 1912 the Observatory purchased a Mack truck, the skeleton of which still reposes in the 100-inch telescope building. Its tires were solid, its springs rudimentary, and its riding qualities were similar to those of a tractor in a plowed field. Not long after the truck went into commission, Abbot arrived for his annual visit to Mount Wilson. He had had wide experience with all the other methods of transportation, and we placed him on board the truck wondering what his reaction would be. About three hours later my telephone rang and after a moment a voice said: "This is Abbot. As I came up the mountain this morning I was reminded of a verse from an old hymn of my childhood, 'Shall I be carried to the skies on flowery beds of ease while others fight to win the prize, or sail o'er stormy seas?' "

We had many other scientific visitors during these early years, among them Evershed of England, Julius of anomalous dispersion fame, and E. F. Nichols, who made some measurements of very long-wave solar radiation with a radiometer. The visit which I remember best, however, was that of Simon Newcomb,† chiefly because of the variety of troubles I experienced in connection with it. The first intimation of his coming was in the form of a telegram sent from a Santa Fé train announcing that he was arriving the following morning at the Santa Anita station to go up Mount Wilson. Hale was in the East and it devolved upon me to make the necessary arrange-

† Director, *Nautical Almanac Office*, U.S. Naval Observatory, 1877–1897.

ments and go up the trail with him. By this time the streetcar line had been extended to Sierra Madre, so I telegraphed Newcomb to continue on the train to Los Angeles and come out on the carline. So the next morning I boarded the appropriate Sierra Madre car at the Colorado Street intersection and found Newcomb on board, recognizing him without difficulty although I had never met him previously.

At the foot of the trail we found the two mules, engaged in advance, waiting for us. Newcomb looked at his mule somewhat suspiciously and I had a premonition that his experience with this sagacious animal might have been limited. We succeeded in getting Newcomb into his saddle, however, and started up the trail, my mule leading the way. After we had gone about half a mile we met a pack mule coming down the trail by himself and evidently on his way to his corral and the food he was anticipating. As was often the case after going up the mountain in a packtrain, he had been unloaded and then turned loose to find his own way back home. The mule brushed past and my animal proceeded on his way. A minute or two later, however, I looked back to see how Newcomb was progressing and found that he had disappeared completely. He had been riding with a loose rein and his mule had seized the opportunity to pivot around and follow his comrade back to the delights of the stable. So I returned to the foot of the trail and there found Newcomb in a state of great excitement vowing that the trip was too dangerous; and it required the united persuasion of everyone at the corral to get him back into the saddle again. This time we succeeded in reaching the top of the mountain, but the journey was a hectic one. The day was warm and Newcomb proceeded to open an umbrella to protect his head from the sun. This did not improve the disposition of the mule who shied every time the umbrella touched a bit of brush along the way. So our progress was in a series of jerks as the mule shied, and these worried Newcomb as he looked down the steep canyon walls. In addition, he seemed to anticipate meeting in such wild surroundings every sort of creature from a centipede to a brown bear. All these factors tended to emphasize certain characteristics of Newcomb wherein he differed from the Christian saints, and by the time we reached the Monastery he was in a state of mind to make a dinner upon a menu of wire nails.

At the Observatory the only things which interested Newcomb to any extent were the Riefler clocks in the laboratory and the Snow telescope building. These he examined closely. The coelostat and concave mirror combination he dismissed briefly as not forming a real telescope, and the spectrographs and spectroheliograph were not merely a closed book to his mind but one which ought to remain closed. In fact, Newcomb's attitude toward the spectrum resembled closely that of Burnham whose famous remark to Hough, "Gale showed me the soda lines once at the Ryerson Laboratory but I didn't think much of them," has become a classic in astronomical literature.

The following day Newcomb and I returned to the valley, the plan being to use the same mules which had brought us up the mountain and had been kept at the hotel for this purpose. As soon as Newcomb sighted the animals, however, he announced that he would try walking for a short distance. So he started out and I followed, riding one mule and driving the other. Expecting to overtake him at any moment I did not actually do so until we reached the halfway house on the Sierra Madre trail. There I found Newcomb sitting at a table drinking ginger ale out of a bottle and enormously pleased with himself over his ability to cover four miles of the mountain trail on foot. I strongly suspect that his satisfaction was due quite as much to the fact that he had avoided riding his mule for this distance. He was ready by this time, however, to board his animal and we rode the rest of the way without difficulty. When I left him on the streetcar on his way to Los Angeles, he was a vastly more genial Newcomb than I had known previously, doubtless because he was treasuring the thought that he had placed behind him the wild and dangerous expedition to Mount Wilson. . . .

The construction, transportation, and erection of the 60-inch telescope with its building and dome furnished a major problem to the Observatory for a period of nearly two years. The dome and the heavy parts of the mounting were built by the Union Iron Works of San Francisco, and the driving clock and the control mechanisms by an instrument shop in Pasadena. Transportation was a very serious difficulty. The new trail had to be widened into a road, and the work had to be done almost wholly by hand, with mule-drawn scrapers and plows to move the earth. At many places bags filled with sand were used to build retaining walls to support the road. Actually the construction of the toll road, as it was known in later years, took place in two stages: the first widening was to provide for the transportation of the 60-inch telescope; and the second and final widening about ten years later for that of the 100-inch telescope. At the time of the second widening some use could be made of mechanical equipment, such as power shovels and drills, although hand labor was still necessary for much of the work.

For the transportation of the heavy parts of the telescope the Observatory purchased a remarkable type of truck, only a few units of which were ever manufactured. Its power plant consisted of a gasoline engine which operated a generator which in turn furnished electric current to four motors installed in the wheels of the truck. The independent operation of the wheels was most useful on the sharp turns of the narrow mountain road, but the power proved inadequate for the steep grades. Finally the mechanical power was supplemented by the addition of mulepower, and a pair of these animals formed a regular part of the equipment when the heaviest loads were being taken up the mountain. Since the rate of progress of this caravan was very slow, a trip to the summit could not always be made in one day, and camping equipment was taken along as well. The picturesque character of this complex outfit had to be seen to be fully appreciated. . . .

The completion of the 60-inch telescope brought a considerable increase in the staff of the Observatory. Seares came to organize the computational work and to begin his admirable series of photometric investigations. Babcock also came, of whom Hale once said: "What the Observatory needs is half a dozen Babcocks." King began his fundamental studies with the electric furnace in the Pasadena laboratory, and the activities of the Observatory expanded in many directions. St. John, whose genial personality and ability in research were such an asset to the Observatory for nearly thirty years, was added to the staff of solar observers in 1909; and John Anderson with his extensive experience at the Johns Hopkins University and his profound knowledge of applied optics came shortly afterward to undertake plans for the ruling machine. Observations with the 60-inch telescope were carried on at first mainly by Ritchey, Babcock, Adams, and Pease, who had assisted previously in the design and construction of the telescope in the Pasadena shop. Van Rijn and van Maanen came in the summer of 1911 as volunteer assistants, and van Maanen remained as a member of the staff to begin his association of thirty-five years with the Observatory. Arnold Kohlschütter of the Hamburg Observatory also came as a temporary visitor but continued as an assistant on the staff until the outbreak of the first World War in 1914. Störmer, Hertzsprung, and others carried on investigations for a few months on the mountain, and Professor Kapteyn made the first of the successive visits which were of such great value in planning the program of work of the 60-inch telescope.

This was a period of intense activity in many fields. Hale had just discovered the Zeeman effect in sunspots, and measurements of field strengths in spots and of line-separations in laboratory sources were being carried on vigorously; studies of the rotation of the sun, of convection currents in spots, and of general circulations in the sun's atmosphere were in progress;

the effects of absolute magnitude upon stellar spectra were being investigated, and the discovery of the spectroscopic method of determining the distances of stars was close at hand; and the magnificent direct photographs with the 60-inch telescope were beginning to open the rich field of nebular study. Moreover, most of the members of the staff were in the prime of life and, for example, thought little of walking up the mountain in an afternoon, adjusting or changing their instruments, and working throughout the following night. The only cloud on the horizon was Hale's illness, which necessitated his absence from the Observatory for more than a year. He never recovered his health completely in later years and could make relatively few trips to the mountain, which was never the same without him.

The year 1910 was notable for the visit of Mr. Carnegie, the meeting of the International Solar Union on Mount Wilson, and the completion of the 150-foot tower telescope. Mr. Carnegie's visit took place in March and the party included Mrs. Carnegie and their daughter Margaret, together with Mr. and Mrs. Hale and Margaret Hale. The two young girls were of nearly the same age and their activity added greatly to everybody's enjoyment. Transportation was by horse and carriage up the toll road, still something of an adventure in those days. The weather was cold and I remember that the night observers contributed several fur coats to the comfort of some members of the party. It snowed during the night which was spent on the mountain, but fortunately not enough to block the road on the following day. This visit was the occasion for the construction of the famous "Carnegie box" which ornamented the 60-inch telescope dome for so many years, and was invaluable in the changing of heavy instruments at the Cassegrain focus. Hale had feared that Carnegie would be unable to climb the steps to the floor of the dome and this box was built for the purpose of raising him to the floor by means of a chain hoist. It was made of two-inch planks reinforced at the corners with metal rods to inspire special confidence, and was designed by Ritchey who believed in factors of safety of one hundred or more. Fortunately, Mr. Carnegie was not informed of the plan in advance. The box was placed on the concrete floor near the foot of the stairs, but when Carnegie arrived he took merely a casual glance at it and went up the steps like a boy out of school.

Mr. Carnegie was in excellent form during this visit. He was photographed repeatedly with Hale, always taking the higher position on a slope since he was somewhat sensitive about his height. To a reporter who asked him a question he remarked with a twinkle in his eye, "The protective tariff has outlived its usefulness." He was much interested in the Observatory equipment and asked many questions about the work in progress. That he did not altogether grasp the scientific point of view was evident when on his return to New York he announced that "60,000 new stars had been discovered" with the telescope at Mount Wilson, evidently a reference to Ritchey's photographs of the Hercules cluster. This elicited the rather cynical remark from Moulton of the University of Chicago that one might equally well speak of having "discovered 60,000 new gallons of water" in Lake Michigan. Nevertheless, a great deal could be forgiven, for within a few weeks Mr. Carnegie showed his appreciation in a highly practical and generous way. In a letter to Dr. Woodward, president of the Institution, Carnegie stated that he was adding $10,000,000 to its endowment and expressed the desire that the work at Mount Wilson be pushed forward as rapidly as possible. This had particular reference to the completion of the 100-inch telescope, a project initiated a few years before by Mr. Hooker of Los Angeles with a gift of $45,000 for the purchase of the mirror. The glass disk was available at the time of Mr. Carnegie's visit. . . .

The 100-inch telescope gradually came nearer completion after many delays due to the first World War, and one night, that of November 1, 1917, it was ready for its first test. Hale, whose health was far from satisfactory, came up from the valley bringing with him the English poet Alfred Noyes, who was in California at the time. We were hopeful but none too certain about the optical performance of the telescope, since Ritchey insisted that the large mirror changed its figure with position, and had what he called "a strong and a weak diameter." Although we were reasonably sure that the effect he had observed was due to conditions in the testing hall rather than to the mirror itself, the nonhomogeneous interior of the disk was an uncertain factor. Soon after dark the telescope was swung over to the eastward and set on the planet Jupiter, and we had our first look through the great instrument. The sight appalled us, for instead of a single image we had six or seven partially overlapping images irregularly spaced and filling much of the eyepiece. It appeared as if the surface of the mirror had been distorted into a number of facets, each of which was contributing its own image. On inquiry we found that the dome had been open throughout the day while the workmen were busy with parts of the mounting, and it even seemed probable that the sun had shone, if not upon the mirror itself, at least upon the cover above it. This information cheered us up somewhat, although we were still much depressed, for our experience with the heating of mirrors by the sun had not prepared us for distortions on such a scale, and it seemed hardly possible that this could be the complete explanation. For two or three hours, however, we remained in the dome, now and then taking a look into the eyepiece. The image was changing, but the improvement was not great. To add to the gloom, news of the great disaster to the Italian army at Caporetto had just arrived, and I remember our sitting around on the floor of the dome speculating on whether Italy was completely out of the war.

Finally we decided to go back to the Monastery, but Hale and I made an engagement to meet at three o'clock in the morning at the telescope building. I doubt whether either of us slept in the interval, for we both arrived ahead of time. Jupiter was out of reach in the west, so we turned northward to the bright star Vega. With his first glimpse Hale's depression vanished: the mirror had resumed its normal figure during the long cool hours of the night, and the image of the star stood out in the eyepiece as a small sharp point of light, almost dazzling in its brilliancy. The success of this great instrument was fully assured.

In the morning we reported our experiences to Alfred Noyes. His instinct for the dramatic seized upon them, and they have become a part of his great picture of the progress of astronomical science throughout the ages, "The Watchers of the Sky. . . ."

In looking backward over the earlier years of the history of the Observatory, I am impressed more and more by the greatness of the figure of Hale, without whom it is doubtful whether the Observatory, if established at all, could ever have attained its high position in the field of astronomy. For the early years were difficult years filled with uncertainties, and it took great courage to start at the very beginning and, as time passed by, to face the responsibility for sponsoring such costly and sometimes unpredictable instruments as the 60-inch, 100-inch, and 200-inch telescopes. A lesser man with less creative imagination would have been satisfied with more conservative equipment which he might himself use, but Hale planned for a future in which he had long realized he could take no active part. We can rejoice that he lived to see the success of these great plans assured. As one who was associated with him from the beginning, I can but express my feeling of the greatness of the privilege allowed me in our long period of association together.

VI
EPILOGUE

*A*lfred D. Hershey, Director, Genetics Research Unit, Carnegie *Institution, is internationally known for his researches in genetics, especially the genetics of bacteriophage. Dr. Hershey has been associated with the Carnegie Institution since 1950. A native of Michigan, he was graduated from Michigan State College in 1930, and received the Ph.D. in chemistry there in 1934. He was associated with the Department of Bacteriology, Washington University School of Medicine, St. Louis, from 1934 until 1950, when he joined the Carnegie Institution's Department of Genetics.*

Alfred D. Hershey

FAITH AND THE SCIENTIFIC ENDEAVOR

From the Introduction to the annual report of the Genetics Research Unit, Carnegie Institution *Year Book 65*.

The enduring goal of scientific endeavor, as of all human enterprise, I imagine, is to achieve an intelligible view of the universe. One of the great discoveries of modern science is that its goal cannot be achieved piecemeal, certainly not by the accumulation of facts. To understand a phenomenon is to understand a category of phenomena or it is nothing. Understanding is reached through creative acts.

The universe presents an infinite number of phenomena. The faith of the scientist, if he has faith, is that these can be reduced to a finite number of categories. Even so he tends to consider the path toward his goal as endless. Not too discontentedly, either, because human history is replete with glorious paths, not goals achieved.

To speak of goals at all is to speak in unscientific terms. One cannot measure progress toward the goal of understanding. Various peoples at various times have thought they had an intelligible view of the universe and, so thinking, had in fact. Most of us today, in spite of much talk about contemporary spiritual malaise, are complacent enough intellectually. If understanding is reached through creative acts, they are partly acts of faith.

These are large questions. They are pondered by professional thinkers, who evidently believe in the power of

329

abstract thought. If that power is efficacious, it behooves the scientist to exercise it now and then when his experiments flag. Otherwise he risks failing a personal goal: to see his work in selfless perspective.

DISCARD
BENNY COLLEGE LIBRARY